1918

Bicycle Accessories

AND REPAIR SUPPLIES

BICYCLES
MOTORCYCLE ACCESSORIES
MOTOR ACCESSORIES

Since the above photograph was taken a large addition has been made to this Weston plant. Total floor space is 110,000 square feet.

by

Canada Cycle and Motor Company

LIMITED

MONTREAL TORONTO **WESTON** WINNIPEG VANCOUVER

 ISBN #978-1-935700-52-4

Canada Cycle and Motor Co., Limited, Vancouver, B.C.
Area, 9,000 square feet.

Canada Cycle and Motor Co., Limited, Montreal
Total of 15,209 square feet occupied by Montreal Branch.

WHEN ORDERING

PLEASE write plainly, giving catalogue numbers and descriptions. Always state how you wish us to ship, either by freight, express or parcel post. Unless these instructions are given we reserve the right to use our own judgment in making shipment.

PRICES

F.O.B. Factory or nearest Branch. All prices in this catalogue are based on delivery as follows:

Provinces of Quebec, New Brunswick, Nova Scotia and Prince Edward Island, **F.O.B. Montreal.**

Province of Ontario, as far West as Fort William, **F.O.B. Weston, Ont.**

Points in Ontario West of Fort William, also Provinces of Manitoba, Saskatchewan and Alberta, **F.O.B. Winnipeg.**

Province of British Columbia, **F.O.B. Vancouver.**

TERMS

15 days from date of shipment. Strictly net. Interest at 7% charged on all overdue accounts.

Collection by Bank Draft unless otherwise arranged.

All prices are subject to change without notice.

GUARANTEE

We endeavour to represent each article catalogued exactly as it is.

All goods of our own manufacture, or which bear our guarantee, are subject to replacement should they prove defective. We will pay transportation charges one way on defective goods returned for replacement.

BACK ORDERS

We endeavour at all times to keep a stock of every article listed in this catalogue, but in these War times, with shortages of material and labor and delays in transportation, it may occasionally happen that we are temporarily out of stock on certain items.

Any such items which we back order will be sent forward with the next goods, or if there is sufficient weight to make a separate shipment they will be forwarded as soon as received.

Kindly note that we do not prepay transportation charges on back orders.

PACKING CHARGES

Packing cases are charged at cost, and we will gladly credit any cases which are returned to us, charges prepaid.

SHORTAGE AND DAMAGE CLAIMS

We endeavour to pack carefully. Any claims for shortages should be made within five days from receipt of goods.

If shipment is delivered in a damaged condition by the Transportation Company, do not sign for shipment and make a verbal claim, as in this way you give the Transportation Company a clear receipt. Before signing, write on the Transportation Company's receipt that the shipment is in a damaged condition. Then place your claim with the Transportation Company for the amount of damage and we will be glad to render every assistance possible in helping to collect your claim.

FIRST CLASS FREIGHT RATE ON BICYCLE TIRES

Bicycle tires are entitled to first class freight rate, but are frequently charged at 1½ times first class rate. This is because tires other than Bicycle take the higher rate in the classification. Bicycle tires, however, are entitled to a rating of first class.

Please check over freight bills carefully and apply for refund in case of overcharge.

RETURN OF GOODS

Goods properly shipped on bona fide orders will not be accepted for credit, unless arrangement is first made with us for return of same and will be subject to discount unless received in as good condition as when shipped from here.

Goods made to order or ordered specially will not be accepted for credit.

PUT YOUR NAME ON RETURN GOODS

When returning goods, please make sure that your name is plainly marked on the return tag. Also write us the day the goods are returned, giving full particulars. We frequently have a number of shipments of returned goods on hand with nothing whatever to indicate from whom they came.

TELEGRAPH AND TELEPHONE CHARGES

We cannot accept collect telephone or telegraph charges. If through some error of ours a telegram or telephone conversation is necessary we will issue credit for the cost of the message in the regular way.

Canada Cycle and Motor Co., Limited Winnipeg, Manitoba

Area, 10,000 square feet.

WAR CLAUSE

All orders and contracts are accepted only contingent on strikes, accidents and delays of carriers, and subject to change in price or cancellation (with or without notice) depending upon War conditions or causes beyond the control of the Canada Cycle and Motor Company, Limited.

Your Interests are Ours

THE constant aim of this Company is to give the very best service possible.

We are satisfied when everything is right—but not until then.

We value the orders of small dealers, and give them the same careful attention as large orders. A very large proportion of our business is made up of such comparatively small orders.

"The dealer's interests are our interests."

Announcing the 1918 Models of
C. C. M. BICYCLES

ON the following pages we illustrate and describe the various 1918 models in the principal C.C.M. nameplate lines.

We introduce this year two new models, a sturdy Road Racer, Model "T." and a Motorbike type, Model "W."

These are handsomely finished in rich dark green with Gold Sunburst Head and gold stripe, or, as an option, in Road Cart Red with Green Sunburst Head and green stripe.

Other models are finished in rich Glossy Black enamel with the exception of the Red Bird, which is finished in Maroon.

The following additional colors and stripings may be had at a reasonable additional cost:

No. 1—Black with Green and Gold stripe.
No. 2—Black with a Red Head.
No. 3—Maroon only
No. 4—Maroon and Gold stripe.
No. 5—French Grey with Green and Red stripe.
No. 6—French Grey with Blue Sunburst Head, Blue and Gold stripe.
No. 7—French Grey with Red Sunburst Head, Red with Gold stripe.
No. 8—Road Cart Red with Green stripe.
No. 9—Green with a Gold stripe.
No. 10—Green with a Gold Sunburst Head and Gold stripe.

The only difference between Grade "A" and Grade "B" models is in the equipment. The frames and every other part of the bicycle are exactly the same in both models.

When shipping one bicycle we recommend that it go forward by Express, as on this weight the Express charge is usually little, if any, higher than the Freight charge would be, and the Express service of course is much quicker.

Agencies: Each C.C.M. nameplate bicycle is sold only to one Agent in each District. Write for dealers' prices if they are not already represented in your locality.

Guaranteed

Massey Bicycles

Massey Men's Roadster, model 356, 20", 22", 24" or 26" frame. Retail list, net ..

Massey Ladies' Roadster, model 355, 20" or 22" frame. Retail list, net ....

Massey Men's Roadster, model 358 ("B" grade equipment), 20", 22", 24" or 26" frame. Retail list, net ....

Massey Ladies' Roadster, model 357 ("B" grade equipment), 20" or 22" frame. Retail list, net ....

Massey Juvenile, Boys' or Girls', 17¼" frame. Retail list, net ....

Massey Racer, model "T," 20", 22" or 24" frame. Retail list, net ....

Massey Motor Bike, model "W," 22" frame. Retail list, net .

The Express charge on one bicycle is not usually in excess of the Freight charge. We recommend having one bicycle forwarded by Express in all cases.

NOTE.—These bicycles are sold only to one Agent in each District. Write for dealers' prices if they are not being sold in your Locality.

Columbia, Gendron and Ivanhoe Bicycles are also supplied in a similar range of models.

Guaranteed Brantford

Red Bird Bicycles

Red Bird Men's Roadster, model 556, 20", 22", 24" or 26" frame. Retail list, net . .

Red Bird Ladies' Roadster, model 555, 20" or 22" frame. Retail list, net . . . .

Red Bird Men's Roadster, model 558 ("B" grade equipment), 20", 22", 24" or 26" frame. Retail list, net . . . .

Red Bird Ladies' Roadster, model 557 ("B" grade equipment), 20" or 22" frame. Retail list, net . . . .

Red Bird Racer, model "T," 20", 22" or 24" frame. Retail list, net . . . .

Red Bird Motor Bike, model "W," 22" frame. Retail list, net . . . .

Red Bird Juvenile Boys' or Girls', 17¼" frame. Retail list, net . . . .

CANADA CYCLE AND MOTOR CO. LIMITED
BRANTFORD
RED BIRD
WESTON, CANADA WESTON, CANADA

The Express charge on one bicycle is not usually in excess of the Freight charge. We recommend having one bicycle forwarded by Express in all cases.

NOTE.—These bicycles are sold only to one Agent in each District. Write for dealers' prices if they are not being sold in your Locality.

Columbia, Gendron and Ivanhoe Bicycles are also supplied in a similar range of models.

Guaranteed

"Perfect" Bicycles

Perfect Men's Roadster, model 656, 20", 22", 24" or 26" frame. Retail list, net

Perfect Ladies' Roadster, model 655, 20" or 22" frame. Retail list, net ...

Perfect Men's Roadster, model 658 ("B" grade equipment), 20", 22", 24" or 26" frame. Retail list, net ....

Perfect Ladies' Roadster, model 657 ("B" grade equipment), 20" or 22" frame. Retail list, net ....

Perfect Juvenile, Boys' or Girls', 17¼" frame. Retail list, net ....

Perfect Motor Bike, model "W." 22" frame. Retail list, net ....

Perfect Racer, model "T." 20", 22" or 24" frame. Retail list, net ....

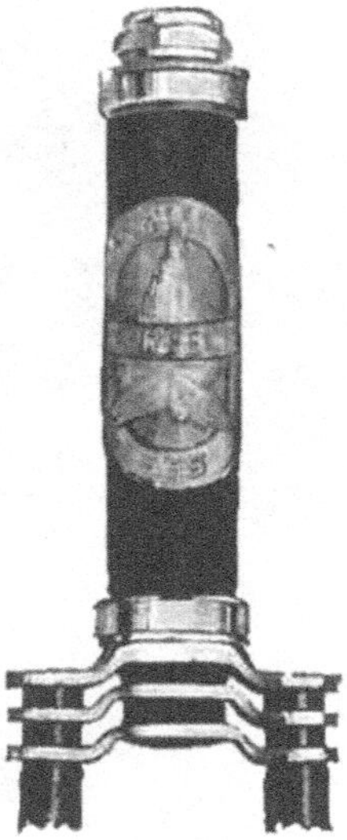

The Express charge on one bicycle is not usually in excess of the Freight charge. We recommend having one bicycle forwarded by Express in all cases.

NOTE.—These bicycles are sold only to one Agent in each District. Write for dealers' prices if they are not being sold in your Locality.

Columbia, Gendron and Ivanhoe Bicycles are also supplied in a similar range of models.

Guaranteed

Cleveland Bicycles

Cleveland Men's Roadster, model 456, 20", 22", 24" or 26" frame. Retail list, net . .

Cleveland Ladies' Roadster, model 455, 20" or 22" frame. Retail list, net . . . .

Cleveland Men's Roadster, model 458 ("B" grade equipment), 20", 22" or 26" frame. Retail list, net . . . .

Cleveland Ladies' Roadster, model 457 ("B" grade equipment), 20" or 22" frame. Retail list, net . . . .

Cleveland Juvenile, Boys' or Girls', 17¼" frame. Retail list, net . . . .

Cleveland Racer, model "T," 20", 22" or 24" frame. Retail list, net . . . .

Cleveland Motor Bike, model "W," 22" frame. Retail list, net . . . .

The Express charge on one bicycle is not usually in excess of the Freight charge. We recommend having one bicycle forwarded by Express in all cases.

NOTE.—These bicycles are sold only to one Agent in each District. Write for dealers' prices if they are not being sold in your Locality.

Columbia, Gendron and Ivanhoe Bicycles are also supplied in a similar range of models.

Repairs Price List

1918 Prices for Enamelling and Nickelling

When Bicycle Complete is Delivered to Us.

	C.C.M. make.	Other makes.
To be enamelled black, including Guards and Rims refinished, all nickelled parts re-plated, supplying new spokes and grips	$24.90	$29.80
Colors other than black, extra	1.40	1.70
Striping extra	1.40	1.70
To be enamelled black and all parts re-plated, with the exception of hubs and spokes	15.80	19.20
To be enamelled black	9.10	11.05
To be enamelled in colors other than black	10.50	12.75
To be enamelled black with striping	10.50	12.75
To be enamelled in color and striped	11.90	14.45

Note.—Any parts which we find badly worn when overhauling bicycles will be replaced by us, unless otherwise advised, and charged for at trade prices.

Handle Bars and Stem to be nickelled and new Grips supplied	$2.30	$2.30
Handle Bars and Stem to be nickelled, without new Grip supplied	1.80	1.80
Cranks nickelled	1.80	1.80
Hubs complete, nickelled and new spokes fitted	7.40	7.40
Front Sprockets	1.40	1.40
Nickelling all parts, less hubs and spokes, when bicycle complete is delivered to us	12.60	15.30
Nickelling all parts when parts only are delivered to us	8.40	10.20
Nickelling all parts (including hubs) and fitting new spokes, when bicycle complete is delivered to us	19.00	22.70

When Frame and Fork or Rims only are Delivered to Us.

Frame and Fork to be enamelled	$2.50	$2.50
Frame and Fork to be enamelled in color other than black	3.50	3.50
Striping on black or colors, extra	1.00	1.00
Fork enamelled	1.00	1.00
Rims scraped and enamelled	2.10	2.60

1918 Prices for Frame and Fork Repairing

	C.C.M. make.	Other makes.
For Frame to be cut down and Frame and Fork enamelled black	$11.90	$14.50
For one Tube in diamond frame and enamelling Frame and Forks black	9.70	10.40
For two Tubes in diamond frame and re-enamelling Frame and Forks black	10.40	12.20
For three Tubes in diamond frame and re-enamelling Frame and Fork black	12.80	14.90
For one rear upper or lower Tube and re-enamelling Frame and Fork black	9.20	11.00
For two rear upper or lower Tubes and re-enamelling Frame and Fork black	10.70	12.70
For three rear upper or lower Tubes and re-enamelling Frame and Fork black	12.90	15.20
For four rear upper or lower Tubes and re-enamelling Frame and Forks black	$14.10	$16.80
For Frame head and re-enamelling Frame and Fork black	9.90	11.70
Saddle Bracket and re-enamelling Frame and Fork black	8.80	10.70
Frame truing ....From $1.40 to	6.00	
For Frame Head upper or lower connection and re-enamelling Frame and Fork black	8.15	10.15
For curved bar in ladies' frame and re-enamelling Frame and Fork black	9.00	10.90
For Frame Head upper and lower connection and re-enamelling Frame and Fork black	9.60	12.10
For crank bracket and re-enamelling Frame and Fork black	12.50	15.00

Note.—The above prices apply when Frame and Fork only are delivered to us. If complete bicycle is sent in, an additional charge of $4.20 for C.C.M. bicycles, and $5.10 for other makes, will be made to cover the cost of tearing down and setting up same.

For new Fork Stem and re-enamelling Fork black	$4.10	$4.90
For new Fork Stem and re-enamelling black and nickelling Crown	4.80	5.80
For new Fork Stem and re-enamelling black and nickelling Crown and tips	6.20	7.40
For splicing Fork Stem	2.80	3.10
For one Fork side and re-enamelling black	4.00	4.70
For one Fork side and re-enamelling black and nickelling Crown	4.70	5.55
For one Fork side and re-enamelling black and nickelling Crown and tips	6.40	7.60
For two Fork sides and re-enamelling black	5.10	5.90
For two Fork sides and re-enamelling black and nickelling Crown	5.30	6.30
For two Fork sides and re-enamelling black and nickelling Crown and tips	7.20	8.40
For one Fork tip and re-enamelling black	2.30	2.80
For one Fork tip and re-enamelling black and nickelling Crown	3.00	3.60
For one Fork tip and re-enamelling black and nickelling Crown and tips	3.70	4.50
For two Fork tips and re-enamelling black	3.20	3.80
For two Fork tips and re-enamelling black and nickelling Crown	3.90	4.70
For two Fork tips and re-enamelling black and re-nickelling Crown and tips	4.60	5.50
For one Fork Crown and re-enamelling black	4.30	5.30
For one Fork Crown and re-enamelling black and nickelling Crown	5.20	6.10
For one Fork Crown and re-enamelling black and nickelling Crown and tips	6.20	7.40
Fork Truing ....From $1.40 to	3.40	

SPECIAL NOTICE

We cannot possibly handle repairs or returns without a letter of instructions, as to take a chance might cause annoyance as well as expense.

"The Dealers' Interests Are Our Interests."

Parts Price List

Always Give Model Number of Bicycle When Ordering

Cleveland, Massey; Brantford Red Bird, Perfect, Gendron, Ivanhoe and Columbia Bicycles using "C.C.M." Hangers, 1916, and Model "T" Hangers, 1917-18 Also Hanger Parts Previous to 1916.

CLEVELAND CRANK BRACKET PARTS.

Models No. 114, 115, 123, 124, 126, 127, 128, 133, 134, 135, 136, 137, 143, 144, 145, 146, 147, 148, 402, 403, 404, 405, 406, 412, 414, 415, 416, 422, 433, 434, 435 and 436, **fitted with C.C.M. Hanger.**

Models No. 455, 456, 457 and 458, **fitted with "T" Hanger.**

Crank Bracket Axle with stationary cone, assembled	Each	$2.60
Crank Bracket Stationary Cone	Each	0.50
Crank Bracket Adjusting Cone	Each	0.50
Crank Bracket Ballcages (No. 322), less balls, 5/16 balls used	Doz.	2.08
Crank Bracket Adjusting Cone Locknut	Each	0.24
Crank Bracket Adjusting Cone Lockwasher	Each	0.04
Cranks, right	Each	1.70
Cranks, left	Each	1.70
Crank Jamb Nut, right	Each	.10
Crank Jamb Nut, left	Each	.10
Sprocket (state what pitch and number of teeth)	Each	2.50
Sprocket Bolt (four used)	Doz.	.48
Sprocket Bolt Nuts (four used)	Doz.	.24
Steel Balls, 5/16 used (7 balls used in each cage)	Gross	2.40
Coupler and Ball Case (solid), used on Models 433, 434, 435, 436	Each	3.00
Ballcase Collar, used on Models 433, 434, 435, 436	Each	.04
Ballcase Cup (press-in style), used on Models from 114 to 422	Each	.50
Right Cup (black finish), used 1917-8 Model T.	Each	.70
Left Cup (nickel-plated finish), 1917-8 Model T.	Each	.70
*Crank Bracket Clamp Bolt	Each	.20
*Crank Bracket Clamp Bolt Nut	Each	.10

*Used on all Models except 455, 456, 457 and 458.

BRANTFORD CRANK BRACKET PARTS.

Models No. 66A, 66B, 77A, 77B, 68A, 68B, 69A, 69B, 70A, 70B, 71A, 504, 506, 514, 516 and 522, **fitted with old Brantford Crank Hangers.**

Crank, right	Each	$3.00
Crank, left	Each	2.50
Crank Bracket Axle	Each	3.00
Crank Bracket Cup, right	Each	1.00
Crank Bracket Cup, left	Each	1.00
Crank Bracket Cup Coupler	Each	.90
Steel Balls, 5/16" used	Gross	2.40
Crank Jamb Screws, left	Each	.20
Crank Jamb Screws, right	Each	.20
Crank Sprocket (state what pitch and number of teeth)	Each	2.50
Crank Sprocket Bolt (four used)	Doz.	.48

Brantford Crank Bracket Parts—(Continued).

Crank Sprocket Bolt Nut (four used)	Doz.	.24
*Crank Bracket Clamp Bolt	Each	.20
*Crank Bracket Clamp Bolt Nut	Each	.10

*Used on all models except 555, 556, 557 and 558.

NOTE.—Models No. 503, 505, 515, 533, 534, 535, 536 **fitted with C.C.M. Hanger.** Models 555, 556, 557 and 558 **fitted with "T" Hanger.**

Prices on these are the same as listed under "Cleveland" on this page.

MASSEY CRANK BRACKET PARTS.

Models A, A1, 43, 44, 46, 53, 56, 64, 83, 84, 85, 86, 87, 302, 303, 304, 305, 312, 314, 316, 322, **fitted with old Massey Crank Hanger.**

Crank, right	Each	$4.00
Crank, left	Each	3.00
Crank Bracket Sleeve	Each	2.80
Crank Bracket Cup, right	Each	1.00
Crank Bracket Cup, left	Each	1.00
Crank Bracket Cup Coupler	Each	.90
Crank Bracket Ballcage, less balls (No. 322), 5/16 balls used	Doz.	2.08
Steel Balls, 5/16"	Gross	2.40
Crank Sprocket (state what pitch and number of teeth)	Each	2.50
Crank Sprocket Bolt (four used)	Doz.	.48
Crank Sprocket Bolt Nut (four used)	Doz.	.24
*Crank Bracket Clamp Bolt	Each	.20
*Crank Bracket Clamp Bolt Nut	Each	.10

*Used on all models except 355, 356, 357 and 358.

NOTE.—Models 305, 306, 315, 333, 334, 335, 336, **fitted with C.C.M. Hanger.** Models 355, 356, 357 and 358, **fitted with "T" Hanger.**

PERFECT CRANK BRACKET PARTS.

Models 60, 61, 62, 63, 64P, 65P, 66P, 67P, 604, 606.

Single-piece Cranks	Each	$5.00
Crank Bracket Cup, right	Each	1.00
Crank Bracket Cup, left	Each	1.00
Crank Bracket Stationary Cone	Each	.50
Crank Bracket Adjusting Cone	Each	.50
Crank Bracket Adjusting Cone Lock Nut	Each	.24
Crank Bracket Adjusting Cone Lock Washer	Each	.04
Crank Bracket Ball Retainer Washer	Each	.06
Steel Balls, 5/16"	Gross	2.40
Crank Sprocket Bolt (four used)	Doz.	.48
Crank Sprocket Bolt Nut (four used)	Doz.	.24
Crank Sprocket (state what pitch and number of teeth)	Each	2.50

NOTE.—Models No. 633, 634, 635, 636, **fitted with C.C.M. Hanger.** Models No. 655, 656, 657, 658, **fitted with "T" Hanger.**

Parts Price List (Continued)

FRAME PARTS.

Cushion Frame (stripped), less Rear, Lower and Upper	Each	$18.00
Cushion Frame (stripped), with Rear Forks assembled	Each	30.00
Rigid Frame (stripped)	Each	20.00
Frame Head	Each	1.50
Frame Head Connection	Each	.30
Front Fork	Each	5.00
Front Fork Stem	Each	.60
Front Fork Crown	Each	.80
Front Fork Side	Each	.80
Frame Front Lower Tube	Each	1.00
Frame Upright Tube	Each	1.00
Frame Top Tube	Each	1.00
Frame Saddle Bracket	Each	.90
Frame Rear Upper Tube	Each	.80
Frame Rear Lower Tube	Each	.80
Frame Rear Tube Brace	Each	.20
Frame Rear End	Each	.20
Frame Rear Tube Plug	Each	.24
Frame Crank Bracket, Malleable	Each	1.80
Chain Adjusting Screw	Each	.08
Chain Adjusting Screw Clip (Yoke)	Each	.08
Chain Adjusting Screw Nut	Each	.02
Frame Head Upper Cup	Each	.36
Frame Head Lower Cup	Each	.36
Frame Head Upper Adjusting Cone	Each	.48
Frame Head Stationary Cone	Each	.20
Frame Head Adjusting Cone Lock Nut	Each	.30
Frame Head Adjusting Cone Lock Nut Washer	Each	.04
Frame Head Ball Cage (less balls)	Each	.10
Frame Head Ball Retainer Washer (used on old models where no Ball Cage was fitted)	Each	.04
Frame Nameplate (for any C.C.M. high grade model)	Each	.60
Frame Nameplate Backs	Each	.20
Frame Nameplate Rivets	Per 100	.40
Frame Seat Post. See Accessories.		
Frame Seat Post Expander for Hygienic	Each	.90
Frame Seat Post Expander Bushing, long	Each	.16
Frame Seat Post Expander Bushing, short	Each	.16
Frame Seat Post Expander Sleeve	Each	.30
Frame Seat Post Expander Nut	Each	.30
Handlebar Post, complete. See page 34.		
Handlebar Post Expanding Bolt. See page 35.		
Handlebar Post Expander. See page 35.		
Handlebar Post (stripped), less Bolt and Nut. See page 35.		
Seat Post Bolt	Each	.14
Seat Post Nut	Each	.16

HYGIENIC FRAME AND HANDLEBAR PARTS.

Rear Upper Fork, complete	Each	$8.50
Rear Upper Fork, stripped	Each	5.50
Rear Upper Arch Crown and Tube	Each	2.00
Rear Lower Fork	Each	6.00
Dust Cap and Head	Each	1.00
Dust Cap Head Screws	Each	.02
Plunger Tube	Each	.50
Cartridge	Each	.40
Brass Bushing	Each	.80
Steel Washer	Each	.04

Hygienic Frame and Handlebar Parts—(Continued)

Leather Washer	Each	$0.02
Springs Nos. 1, 2, 3, 4, 5, 6	Each	.50
Rear Lower Crown, Hygienic	Each	.80
Rear Lower Spring Bolt	Each	.10
Rear Lower Spring Bolt Nut	Each	.04
Rear Lower Flat Spring, short	Each	.20
Rear Lower Flat Spring, long	Each	.20
Rear Lower Flat Spring Rivet	Per doz.	.12
Rear End Connecting Bolt	Each	.06
Rear End Connecting Bolt Nut	Each	.02
Rear End Bushing	Each	.06
Saddle Bracket Binder Bolt	Each	.14
Saddle Bracket Binder Bolt Nut	Each	.16

SILLS' HANDLE BAR PARTS.

Handle Bar Post, stripped	Each	$2.50
Handle Bar Post Clamp	Each	1.00
Handle Bar Post Clamp Bolt	Each	.20
Handle Bar Post Clamp Bolt Nut	Each	.12
Handle Bar Post Hinge Bolt	Each	.10
Handle Bar Post Hinge Bolt Nut	Each	.10
Handle Bar Post Hollow Bolt	Each	.50
Handle Bar Post Plunger	Each	.24
Handle Bar Post Spring	Each	.20
Handle Bar Post Expander	Each	.16

No. 10 HUB PARTS.

Front Hub (stripped) Shell	Each	$0.90
Front Axle	Each	.30
Front Axle Cone	Each	.30
Front Hub Cup (screw-in cups), old style	Each	.20
Front Hub Ball Cage (less balls)	Doz.	1.32
Front Hub Axle Washer	Each	.02
Front Hub Axle Nut	Each	.06
Front Hub Cup Lock Ring (or Nut) (used on old Massey and Brantford front hubs only)	Each	.20
Front Hub Cup Lock Washer (old style)	Each	.04
Front Hub Oiler	Each	.14
Front Hub Dust Shield (old style)	Each	.04
Rear Hub (stripped) Shell	Each	1.50
Rear Hub Axle	Each	.32
Rear Hub Cone	Each	.44
Rear Hub Nut	Each	.06
Rear Hub Washer	Each	.04
Rear Hub Cup (press-in style)	Each	.16
Rear Hub Cup (screw-in style)	Each	.30
Rear Hub Cup Lock Ring (old style)	Each	.20
Rear Hub Cup Lock Washer	Each	.04
Rear Hub Ball Cage (less balls)	Each	.14
Rear Hub Dust Shield (old style)	Each	.12
Rear Hub Oiler	Each	.14
Rear Hub Sprocket	Each	.80
Rear Hub Sprocket Lock Nut	Each	.50
Rear Hub Sprocket Lock Nut Washer	Each	.50

C.C.M. MOTORCYCLE PARTS.— Owing to war conditions we cannot now obtain Motosacoche Engine parts (C.C.M. Motorcycle) from Switzerland. Therefore our stock is not complete. In ordering these parts always send sample, and if we have duplicates we will fill your order.

Advertising Caps and Pennants

IN response to requests from a number of dealers who prefer this form of advertising, we can now supply the excellent lines listed below.

By purchasing these in large quantities and bearing part of the cost ourselves, we can offer them at prices which bring them within the reach of every dealer, and at only a fraction of what they would cost in small quantities.

Advertising Caps

Well made, from very heavy weight of white cotton, with oilcloth sweatband and green-lined visor or peak. Furnished only in bundles of 50, each bundle containing an assortment of sizes suitable for boys, youths and men. Very neatly lettered and designed for Brantford Red Bird, Cleveland, Columbia, Gendron, Ivanhoe, Massey or Perfect nameplates. When ordering **state which nameplate wanted.**

No. 581—Cleveland No. 584—Perfect
No. 582—Massey No. 585—Columbia
No. 583—Red Bird No. 586—Gendron
No. 587—Ivanhoe

Per bundle of 50 assorted sizes. (We do not break bundles.) Price ....Each

Advertising Pennants

Neatly designed on richly colored cotton felts, with name and trade mark of Brantford Red Bird, Cleveland, Columbia, Gendron, Ivanhoe, Massey or Perfect nameplates. When ordering **state which nameplate wanted.** Size 9" x 18" allows pennant to fit inside the frame of all boys' juvenile bicycles, while extra long sizes accommodate them to the largest men's frames.

No. 1633—Cleveland No. 1636—Perfect
No. 1634—Massey No. 1637—Columbia
No. 1635—Red Bird No. 1638—Gendron
No. 1639—Ivanhoe

Packed in bundles of 25 pennants. (We do not break bundles.) Price ....Each

Adjusters, Chain

Banjo

C.C.M.
(Made in Our Own Factory)

No. 200—Banjo (½ gross in box) ....Per doz. pair
No. 221—C.C.M. (less clip) ....Per doz. pair
No. 222—Clip only ....Each

Axles, Hub, Front and Rear

No. 201—Front. Size 5" x 5/16", 24 thread. (25 in box) ....Per doz.
No. 211—Front. Size 5" x 5/16", 26 thread. Fit English front hubs. (25 in box.)....Per doz.
No. 202—Rear. Size 6¼" x ⅜", 24 thread. (25 in box) ....Per doz.

Axle Sets, Hub Complete

No. 204

Complete with drive-fit steel washers on cone, which may be turned down to any size desired.
No. 204—Front Axles, 5" x 5/16" ....Each
No. 205—Rear Axles, 6¼" x ⅜" ....Each
No. 210—Front Axle Set, for English hubs..Each

No. 206

(Made in Our Own Factory)

No. 206 (10)—Front Axle Set, 5/16" x 24 thread. (Small cone diameter ⅞") ....Each

Small cone plated, but not buffed.

No. 207 (10 and 25)—Front Axle Set, 5/16" x 24 thread. (Large cone diameter 1 3/16")..Each

Used on C.C.M. Bicycles. (See note.)

Used on 1917 No. 1047 (25) Front Hub.

No. 225 (25)—Front Axle Set, 5/16" x 24 thread, with new cone No. 224 used on 1918 No. 1047 (25) Front Hub ....Each

No. 208 (10)—Rear Axle Set, ⅜" x 20 thread.Each

Note.—The Small Cones were used on No. 10 Hubs supplied during 1912 and previously.

No. 207 Large Cone Axle Set fits our 1917 No. 25 Front Hub, but 4 washers are used (2 on either side) instead of two (1 on either side), as on No. 10.

Our new No. 1047 (25) Front Hub takes No. 225 Front Axle Set with improved Cone.

Axle Set, No. 10 Extra Parts

(Made in Our Own Factory)

FOR FRONT HUBS.

No. 212—Cones 7/8" diameter, for No. 206 set. Each

No. 213—Cones each 1 3/16" diameter, for No. 207 set .... Each

No. 223—Axles only for No. 206 or 207 set.. Each

No. 215—Axle Nuts for No. 206 or 207 set. Each

No. 216—Axle Washers for No. 206 or 207 set.. .... Each

FOR REAR HUBS.

No. 217—Cones for No. 208 set .... Each

No. 218—Axle for No. 208 set .... Each

No. 219—Axle Nuts for No. 208 set.......... Each

No. 220—Axle Washers for No. 208 set.... Each

Axle Set, No. 25 Extra Parts

(Made in Our Own Factory)

No. 224—Cones, 1 3/16" diameter, for No. 225 set. .... Each

No. 223—Axle for No. 225 set .... Each

No. 215—Axle Nuts for No. 225 set .... Each

No. 216—Axle Washers for No. 225 set .... Each

Bags, Tool

(Made in Canada)

Standard—Size 6½" x 3¼" x 1½". Made from good grade of heavy leather, smooth finish.

No. 250 .... Each

Tool Bags, Complete with Tools

Standard Tool Bag—No. 2908 Monkey Wrench, Hand Pump and No. 1375 Oiler.

No. 251—Complete as above .... Each

Balls, Steel

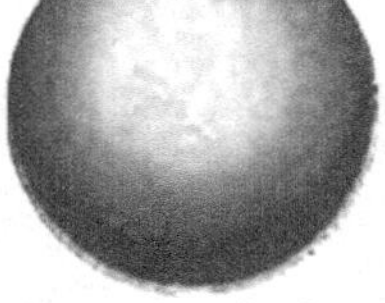

One gross in box.

Sizes 1/16" and 3/32" are not used in Bicycles, but in Phonographs and small mechanical devices.

Sizes 1/8" to 3/8", inclusive, are for use in Bicycles and Motorcycles.

Sizes 13/32" to 1", inclusive, are not used in Bicycles, but in Motorcycles, Motor Cars and large machines.

Diameter.	Price per Gross.
1/16 inch	....
3/32 inch	....

Above are not used in Bicycles.

Diameter.	Price per Gross.
1/8 inch	....
5/32 inch	....
3/16 inch	....
7/32 inch	....
1/4 inch	....
9/32 inch	....
5/16 inch	....
11/32 inch	....
3/8 inch	....

Above are used in Bicycles and Motorcycles.

Diameter.	Price per Gross.
13/32 inch	....
7/16 inch	....
1/2 inch	....
9/16 inch	....
5/8 inch	....
11/16 inch	....
3/4 inch	....
13/16 inch	....
7/8 inch	....
15/16 inch	....
1 inch	....

Above are not used in Bicycles, but in large machines.

Balls Used in C.C.M. Bicycles

1916, 1917 and 1918—(All models the same).

Hanger	3/16"
Pedals	1/8"
Head Fittings	5/32"
Front Hub	1/4"
Hercules Brake	1/8"

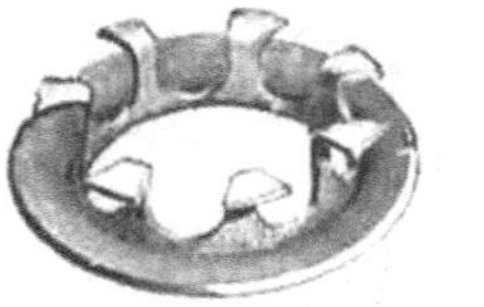

Ball Retainers

BICYCLE

(Less Balls)

Order by Number.

(H 13)—For Hercules C.B. Adj. Cone. Per dozen

(H 14)—For Hercules Cone and Drive Screw. .... Per dozen

(B 115)—For Morrow, large size ..... Per dozen

(B 106)—For Morrow, medium size ... Per dozen

(B 109)—For Morrow, small size ..... Per dozen

(A 16)—For New Departure, large ... Per dozen

(A 20)—For New Departure, small ... Per dozen

(F 56)—Fauber Bracket .... Per dozen

No. 252—For Nos. 10, 25 and 30 Front Hubs, fitted with large cone (diameter 1 3/4") .... Per dozen

No. 253—(90)—For No. 10 Front Hubs, fitted with small cone (diameter, 3/8") .... Per dozen

No. 321—For "Hercules" Single Piece (Model F) Crank Hanger .... Per dozen

No. 322—For C.C.M. 1916 and 1917 Crank Hanger .... Per dozen

NOTE.—No. 322 Ball Retainer used on 1916, 1917 and 1918 (Model T) Crank Hangers.

Ball Retainers

MOTORCYCLE

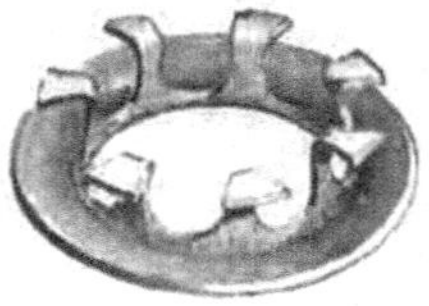

1⅝ ins. No. 60 for Corbin M.C. Hub ..... Doz. $2.52

1 31/32 ins. No. 97 for Corbin M.C. Brake. ... Doz. 2.52

1 31/32 ins. No. 34 for Corbin M.C. Brake ... Doz.

1 13/32 ins. No. 67 for New Departure M.C. Brake. .... Doz.

1 31/32 ins. No. 34 for New Departure M.C. Brake. .... Doz.

1 9/16 ins. No. 30 for New Departure M.C. Front Hub .... Doz.

1 31/32 ins. No. 97 for Miami Cycle & Mfg. Co. M.C. Brake .... Doz.

1 37/64 ins. No. 99 for Miami Cycle & Mfg. Co. Crank Hanger Bracket .... Doz.

1 9/16 ins. No. 54 for Indian M.C. Counter Shaft. .... Doz.

1 13/32 ins. No. 70 for Indian M.C. Adjustable Counter Shaft .... Doz.

2¼ ins. No. 87 for Eclipse Machine Co. M.C. Free Engine Pulley .... Doz.

2⅛ ins. No. 109, used on the Henderson.... Doz.

1⅞ ins. No. 69, used on Reading Standard. .... Doz.

1 31/32 ins. No. 34, used on Thor M.C. Brake. .... Doz.

Bells

For HORNS, see page 37

OUR line of Bells has been carefully selected. They are complete in grade, style and design. They are manufactured of selected metal, producing loud, penetrating and musical tones. The better grades are all manufactured from special bell metal—not steel. **Each bell is packed in a neat cardboard carton.**

No. 256.

Electric Stroke—Union Jack Flag Top, in colors. Flanged gong. Full nickel finish. Screw clamp.

No. 256 (184)—2" gong. ~~Per dozen~~ (12 in box) ....

No. 259.

Double Stroke—Chiming—Wrought metal gong. Full nickel finish. Screw clamp.

No. 259 (104)—Echo, 2½" gong. Per dozen (12 in box)

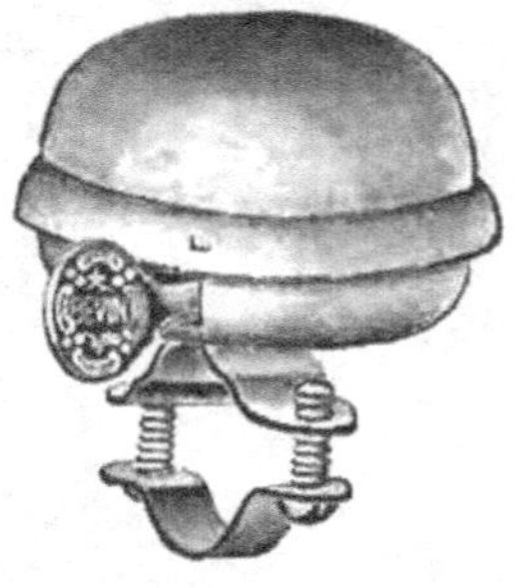

No. 254.

Electric Stroke—Wrought metal flanged gong. Full nickel finish. Screw clamp.

No. 254 (810)—1⅝" gong. ~~Per dozen~~ (12 in box) ....

No. 255 (710)—2" gong. Per ~~dozen~~ (12 in box) ....

No. 257 (610)—2¼" gong. ~~Per dozen~~ (12 in box)

Bells (Continued)

No. 258.

No. 258 (56)—Beaded Push Button. 2¼" beaded gong, continuous ringing. Flag on top in colored enamel, balance of bell nickel finished. Per dozen (12 in box) ....

No. 323. Illustrating both sides. No. 323.

Double Dome, Electric Stroke—This bell has a "Union Jack" on one side and a "Boy Scout" on the other, in colors. Diameter of gong 2 inches. Screw clamp. A good patriotic line.

No. 323 Per dozen (12 in box) ....

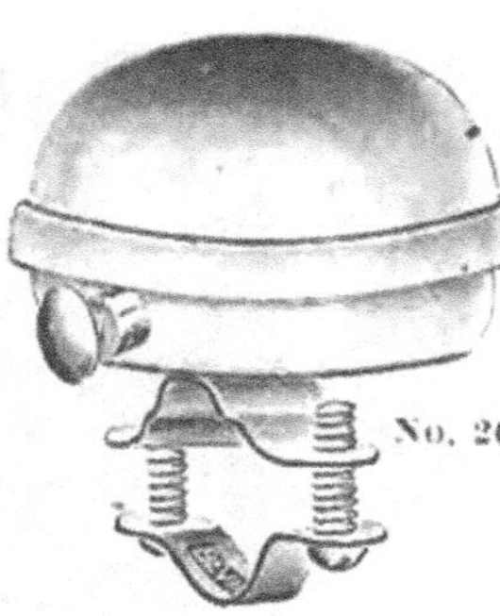

No. 260

Push Button—Continuous Ringing Revolving flanged top. Wrought metal gong. Full nickel finish, screw clamp.

No. 260 (110)—2" gong. Per dozen (12 in box) ....

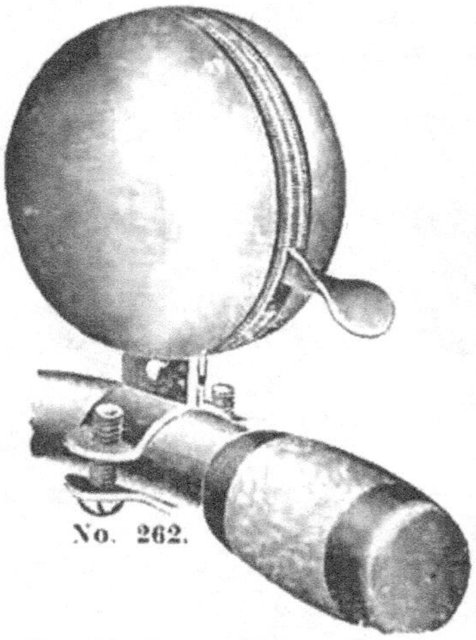

No. 262.

Double Dome Chime, Tandem Gongs—Rotary Movement. Full nickel finish. Screw clamp.

No. 262 (8-0)—2½" gong. Per dozen (12 in box).

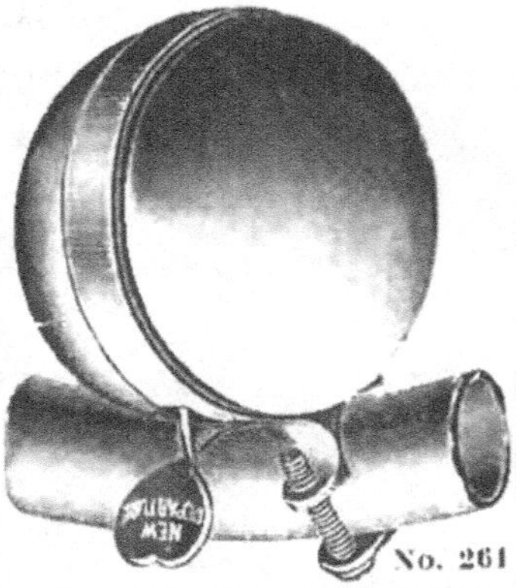

No. 261

No. 261—Cathedral Chime—A high grade chime bell, very clear and musical, made of special selected bell metal, heavily nickel plated and highly finished. Revolving domes 2¼". Screw clamp.

No. 261 (35C)—Per dozen (12 in box)..

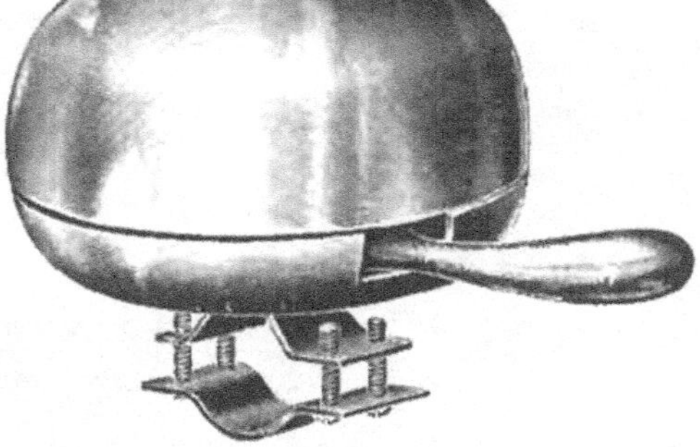

No. 331—Bronko.

No. 331—Bronko—Has a particularly sharp, clear, penetrating note. Full nickel, 3" gong. Large finger ringing lever, easily operated. Double screw clamp. Full 1" size. (One in carton.) Per dozen (12 in box) ......

Bell Assortment "R"

A Very Convenient Assortment for Dealers.

Description—2¼" Electric Stroke, cast metal gongs, nickeled or oxidized, with designs enameled in various colors.

Packed in box of 1 dozen, containing two each of the designs illustrated.

NOTE.—Orders for smaller quantities than one dozen of this assortment will be filled from broken lots, regardless of top design.

No. 265—Per dozen (12 in box) ....

Good Luck Bell Assortment

Made up of three Flag Top, three Boy Scout design, two Four Leaf Clover design, two Race Horse design, and one Lucky pattern design. Bells are 2¼" Electric Stroke, cast metal gongs, flat top type similar to No. 256, and each bell is nicely enameled. Each bell packed in box, and the 12 assorted in one carton.

No. 332—Per dozen (12 in box) ..

Bells, Tire

"Monarch."

"Monarch"—This is an extra well-made Bell. When cable is drawn tight, the band in the centre of the bell comes in contact with the tire, producing a clear, penetrating tone. 3" diameter. Nickel plated.

No. 264—Price (1 in box) ....Each

No. 333.

Side Tire Bell—Since the introduction of extension front mud guards, a great demand has been created for a **side tire bell** which can be fixed to the front fork side and operated by coming in contact with the tire. The ordinary style of tire bell will not fit a bicycle with a front extension guard. This Bell is designed for this purpose. Gongs are 3" in diameter, and the entire bell is nicely nickel plated. Supplied with chain complete.

No. 333—One in box ....Each

Bolts and Nuts

(Assembled)

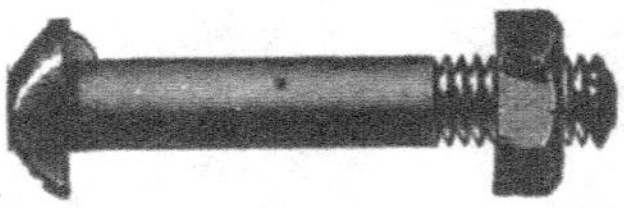

Handle Bar and Seat Post Clamp Bolts and Nuts—These bolts and nuts will give perfect satisfaction. They are accurately threaded and will not strip. Heads plated and buffed.

	Diameter.	Length.		Diameter.	Length
No. 294—	$\frac{5}{16}$ Rd....	1¼"	**No. 298**—	⅜ Rd....	1½"
No. 295—	$\frac{5}{16}$ Rd....	1½"	**No. 299**—	⅜ Rd....	1¾"
No. 296—	$\frac{5}{16}$ Rd....	1¾"	**No. 300**—	⅜ Rd....	2 "
No. 297—	$\frac{5}{16}$ Rd....	2 "			

Price, dozen (1 gross in box) ....Dozen

Bolts and Nuts, Pedal, Sprocket and Toe Clip

(All in packages of 1 Gross each).

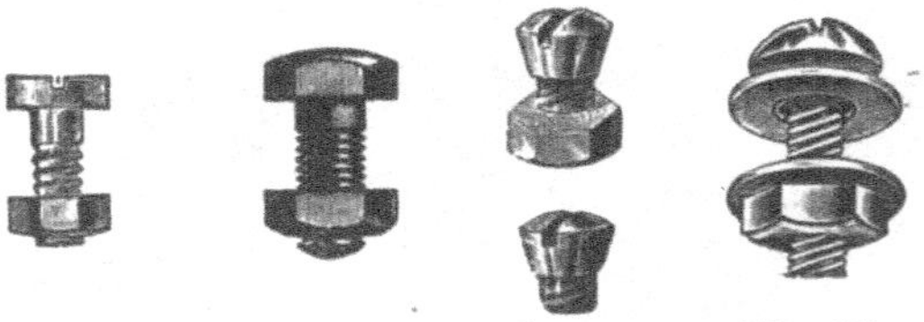

Pedal. **Sprocket.** **Fauber.** **Toe Clip.**

Pedal—These Pedal Bolts and Nuts are nicely nickelled and turned from bar stock steel. The bolts are carefully and accurately threaded, and the nuts fit perfectly.

Sprocket—Our line of Sprocket Bolts and Nuts is manufactured from the highest grade machinery steel, nicely finished and nickel plated. Supplied in $\frac{3}{16}$" and ¼" with 24 threads.

Fauber—We have added a line of short and long Fauber Bevel Head Sprocket Bolts and Nuts, perfect in size and shape, correct in form and finish, complete with nuts.

Toe Clips—These are manufactured from No. 8 wire, and are ⅝" long. They are fitted with two washers, and will hold Toe Clip rigid.

No. 325—Rubber Pedal Bolt and Nut, C.C.M. Per dozen ....

No. 302—($\frac{3}{16}$"). Sprocket Bolts and Nuts. Per dozen ....

No. 303—(¼"). Sprocket Bolts and Nuts. Per dozen ....

No. 304—Sprocket Bolts only, Fauber, short. Per dozen ....

No. 305—Sprocket Bolts and Nuts. Fauber, long. Per dozen ....

No. 324—Sprocket Bolts and Nuts, for all C.C.M. Bicycles. Per dozen ....

No. 306—Toe Clip Bolts and Nuts. Per 100....

Brakes, Improved Hercules Coaster Brake

THIS well known and popular brake has been still further perfected from time to time, and is now giving greater satisfaction than ever before.

The absence of side arm is appreciated by every dealer and repair man, as it enables the brake to be very easily and quickly removed from and replaced in the frame.

Its smooth powerful braking action, and the absence of any slipping when engaging the drive after coasting, appeals strongly to the cyclist.

Other strong features of the Hercules Brake are its extreme lightness, its comparatively small size, and the small number of parts employed in its construction.

Altogether, it is a brake which may be specially recommended and sold with the assurance that it will give the maximum of service and satisfaction and the minimum of trouble.

Guarantee—We give an absolute and unqualified guarantee on the Hercules Brake as to efficiency, material and workmanship, and if it is not all we claim for it, we will replace it at our own expense, if returned to us within one year from date of purchase.

"THE LITTLE FELLOW WITH THE BIG GRIP."

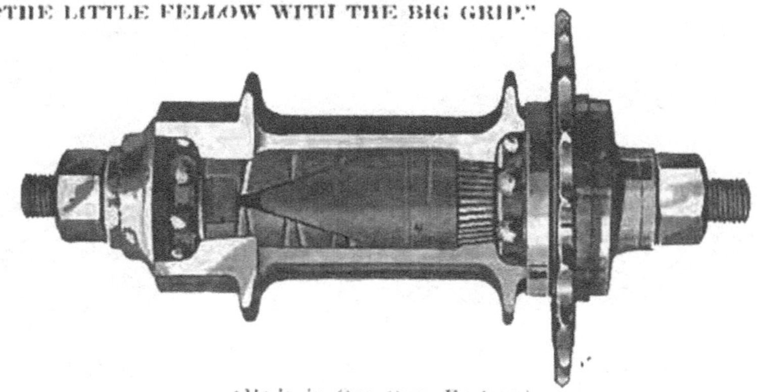

(Made in Our Own Factory)

No. 308 —Price with Sprocket. Each .... [illegible]

Note—Specify Sprocket required when ordering.

Brake Lubrication

For best results from this Brake, use **C.C.M. Hercules Brake Oil**, a heavy bodied oil, especially adapted for this purpose. **See page 46.**

(Made in Our Own Factory)

No. 483. Shim of Hercules Brake.

No. H- 1 —Hub Shell. Each
No. H- 2 —Adjusting Cone .... Each
No. H- 3 —Driving Worm .... Each
No. H- 4 —Driving Clutch .... Each
*__No. H- 5__ —Clutch Drag .... Each
*__No. H- 6__ —Brake Spool .... Each
No. H- 7 —Brake Wedge .... Each

*Assembled together.

No. H- 7½ —Wedge Shim. .... Per dozen
No. H- 8 —Brake Sleeve. .... Each
No. H- 9 —Stationary Cone. .... Each
No. H-10 —Axle .... Each
No. H-11 —Sprocket Lock Nut .... Each
No. H-12 —Sprocket 1" pitch, 3/16" wide, 7 to 10 tooth .... Each
½" pitch, ⅛" or 3/16" wide, 14 to 20 tooth ..... Each

No. H-13 —Small Ball Retainers, less balls ... Each
No. H-14 —Large Ball Retainers, less balls ... Each
No. H-15 —Chain Line Ring. .... Each
No. H-16 —Axle Nut. Each
No. H-17 —Axle Nut Washer. .... Each
No. H-19 —Dust Shield. .... Each
No. H-20 —Brake Oiler. .... Each

Brake Parts, Morrow

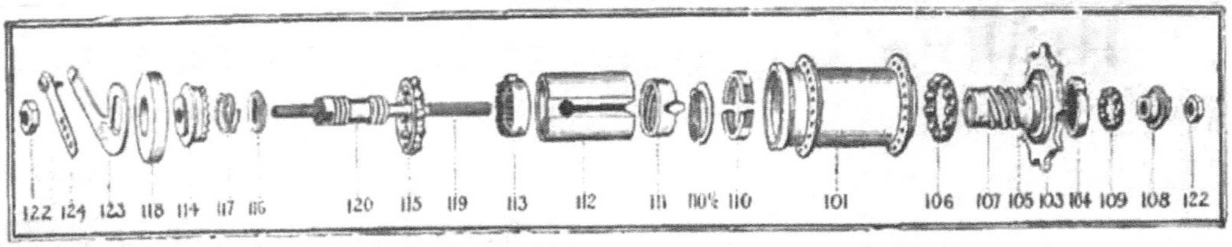

Cannot guarantee to supply all parts shown. Prices on application.

BRAKE PARTS, EADIE.—Prices supplied on application. We cannot guarantee prompt delivery owing to difficulty in securing from England.

- **32R**—Shell, with cups and serrated ring.
- **33R**—Friction Plate
- **34R**—Brake Spring and phosphor bronze ring
- **35R**—Brake Cone and Lever.
- **36R**—Brake Spring Lever.
- **37R**—Clutch Nuts.
- **(A)37R**—New Positive Drive Clutch Nut.
- **38R**—Driving Screw.
- **39R**—Chain Ring (State size.)
- **40R**—R.H. Cups.
- **(B)40R**—R. H. Cups, New Style, with teeth to take New Style clutch.
- **41R**—L.H. Cups.
- **42R**—Chain Stay Clips, complete.
- **43R**—Spindle only
- **44R**—Lock Ring.
- **45R**—Lock Nuts, with oil hole cover.
- **46R**—Ball Retainers, with balls, large.
- **47R**—Ball Retainers, with balls, small.
- **48R**—Clutch Nut Spring.
- **49R**—Adjusting Cone.
- **50R**—Clutch Nut Spring Screw
- **52R**—Split Collar.
- **53R**—Split Collar
- **54R**—Fixing Nut.
- **56R**—Fixing Nut.
- **55R**—Spindle Collar.
- **57R**—Chain Stay Clip Nut.
- **58R**—Chain Stay Clip Screw.
- **59R**—Spindle Nuts.
- **60R**—Lubricator.

Brake Parts
New Departure

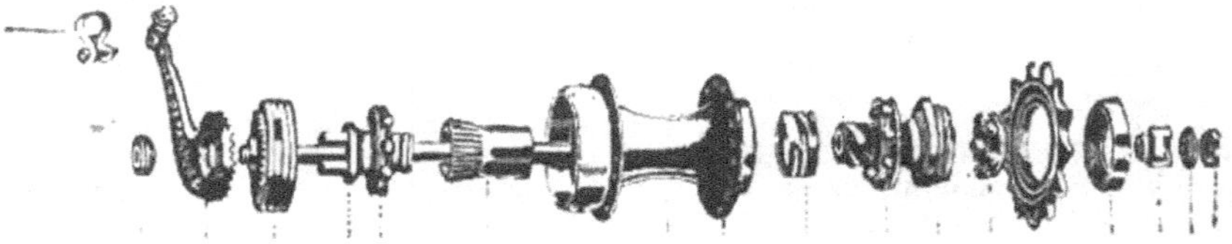

A 11 · A 15 · A 10 · AA 9 8 · AA 6 16 · A 4 · A 3 · A 1 · A 12 · A 16 · A 2 · A 20 · A 17 · A 5 · A 7 · AA 14 13

- **A- 1**—Hub Shell .... Each
- **A- 2**—Drive Screw ...... Each
- **A- 3**—Clutch Sleeve ..... Each
- **A- 4**—Axle .... Each
- **A- 5**—Sprocket Lock Nut... .... Each
- **A- 6**—Clutch .... Each
- **A- 7**—Axle Cone .... Each
- **A- 8**—Brake Box, English.. .... Each (See note.)
- **A- 8**—American .... Each
- **A- 9**—Brake Cover ..... Each
- **A-10**—Brake Arm ..... Each
- **A-11**—Brake Arm Clip. Each
- **A-12**—Transfer Spring.. Each
- **A-13**—Axle Nut .... Each
- **A-14**—Axle Washers.... Each
- **A-15**—Convex Washers. Each
- **A-16**—Ballcage (large) .... Per doz.
- **A-17**—Sprocket ...... Each
- **A-20**—Ballcage (small) .... Per doz.

Note.—State whether A-8 Brake Box is required with short pin (American Brake), or long pin (English Brake).

We cannot supply A-8 English during continuation of war. But A-8 American may be used instead, provided A-9 and A-10 American are also used.

Brake Tool

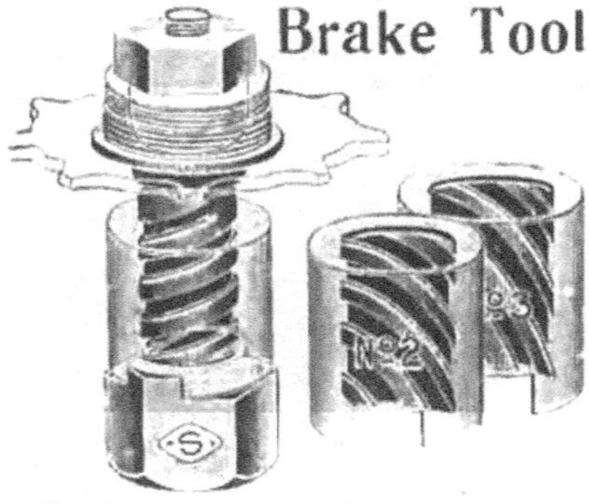

Brake Tool—Stevens' Clutch. A compact Coaster Brake Clutch Tool, for use in vise. Will remove the tightest Sprockets.

Supplied with three special tools that take Hercules, New Departure and American Morrow Driving Screws.

Packed complete in small wooden box.

No. 334 .... Each

Brazing Compounds

- **No. 1157—Borax,** powdered, per lb. ....
- **No. 1158—Boracic Acid,** per lb. ....
- **No. 1156—Spelter,** fine ground, per lb. ....
- **No. 1159—Spelter Wire,** per lb. ....
- **No. 1247—Superior Brazing Compound** (1 pound cans) .... Each

Blow Pipes
Hot Blast, Gas Brazing

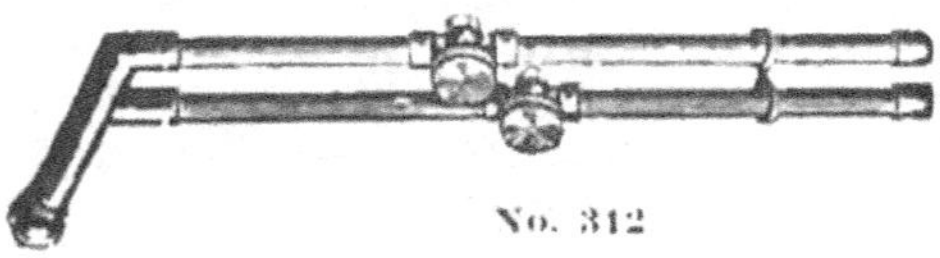

No. 312

Made of ½" brass tubing, carefully fitted. Stop Cocks have pin handle instead of the knurled screw shown in cut. Length 14".

No. 312—Price .... Each

Braziers
Sterling

No. 1447 Brazier, 10 gal. tank with pump) ...... Each

These braziers are fitted with improved hydro carbon burners adjustable to any angle. The tank is made of boiler steel, galvanized, tested to 150 pounds pressure. Gasoline is the fuel used. Gauge, fire bricks and full directions how to operate accompany each brazier.

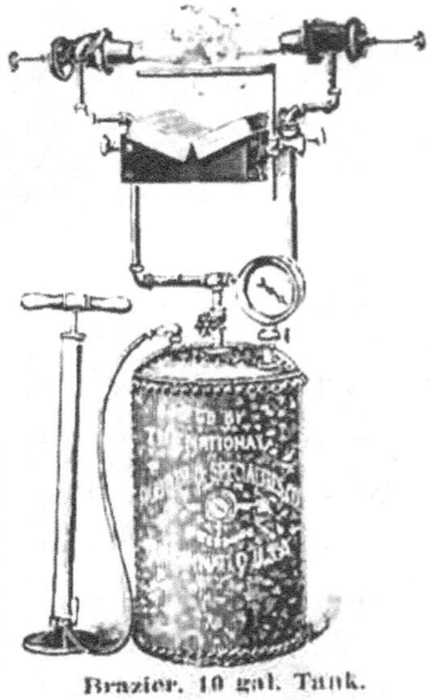

Brazier, 10 gal. Tank.

Prices Subject to Change Without Notice.

Brushes

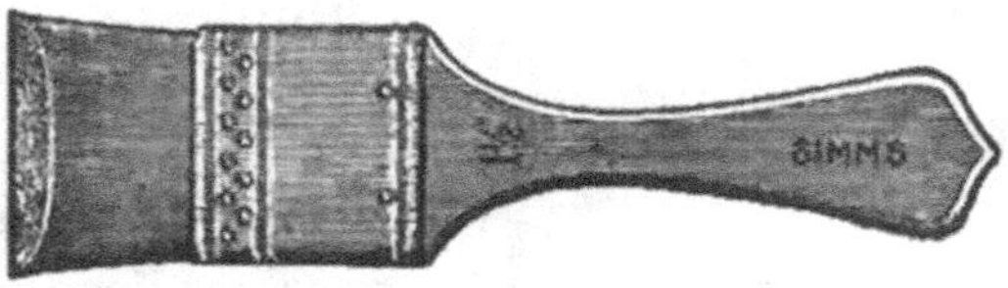

(Made in Canada)

Enamelling.—These brushes are of the very finest quality of hair, cement set, and will apply the enamel to the frame very freely, leaving a perfectly smooth and even surface.

No. 313—Fitch Hair, ½". Per dozen (12 in box)

No. 314—Fitch Hair, 1". Per dozen (12 in box)

No. 315—Fitch Hair, 1½". Per dozen (12 in box)

(Made in Canada)

Bicycle—No. 317—Single strand, 5" long. Length, including handle, 14".

Per dozen (½ gross in carton) ....

(Made in Canada)

Steel Wire—No. 318—Furnished in one size only, 6" x ¾". Length, including handle, 15". Made from the very best materials.

Per dozen (12 in package) ....

Buffers

No. 319—Frame Buffers—These are used to prevent the handlebar from injuring the enamel. Spring style grey rubber.

Per dozen (24 in box) ....

Belts, Motorcycle

Half-inch Round

Chrome tanned, rawhide, fitted with improved fastener, suitable for C.C.M. Motorcycle. Length 7 feet. Complete instructions with each Belt.

No. 266—Round. Price .... Each

Belts, Spartan Motorcycle

Spartan Belts are made of leather, especially adapted for Motorcycle use. Fastened together with cement, which is absolutely water and heat proof, and securely riveted. With a Free Engine Pulley, the "V" Belt is a most satisfactory and popular transmission. "V" Belt, 28 degree angle.

No. 269— ¾" x 4 Ply for single cylinder machines......Price, per foot
No. 270— ⅞" x 4 Ply for single cylinder machines......Price, per foot
No. 271—1" x 4 Ply for single cylinder machines......Price, per foot
No. 272—1⅛" x 4 Ply for single cylinder machines......Price, per foot
No. 273—1¼" x 4 Ply for single cylinder machines......Price, per foot
No. 281—1⅛" x 5 Ply Giant for twin cylinder machines. Price, per foot
No. 282—1¼" x 5 Ply Giant for twin cylinder machines. Price, per foot
No. 283—1½" x 5 Ply Giant for twin cylinder machines. Price, per foot

"V" Shape

Sizes, Flat Belt, 9 ft. or Shorter.

No. 274—1½" ....Price, each
No. 275—1⅝" ....Price, each
No. 276—1¾" x 7' 11" (No. **276A** 8' 2") (No. **276B** 8' 7")..... ....Price, each
No. 277—1⅞" ....Price, each
No. 278—2" x 8' 4" (No. **278A** 8' 7½") ....Price, each
No. 279—2⅛" ....Price, each
No. 280—2¼" ....Price, each

Give length when ordering. Supplied Endless.

Flat.

Bicycle Flags and Holder

Flags of The Allies

Seven silk flags, supplied with black enamelled holder for attaching to handlebars.

No. 759 (3)—Size of each flag 3" x 2" ....Per set

No. 760 (4)—Size of each flag 4¾" x 3" ....Per set

Belt Dressing

Motorcycle

A Belt Food prepared for Motorcycle Belts. It will positively lengthen the life of a belt, and produce a uniform drive. Treat your belt as you would your motor; clean belt often and apply Belt Dressing.

No. 284—Price per dozen Tubes . . . .

Belt Drill

for "V" Belts

No. 654.

Cuts a clean straight hole for sizes ¾", ⅞", 1". Thumb Screw nickelplated

No. 654—One in box. . Each

Belt Fastener

for Motorcycle "V" Belts

We recommend this Fastener for all sizes of Spartan "V" Belts.

Instructions.—To attach, remove rivet from belt by cutting off end of rivet with a cold chisel. Use the rivet holes for the screws. If belt is too thick, shave "block" so that the fasteners will fit snugly.

To detach, simply open the oblong washer on chain link, which is split at one end.

No. 285—Price . . . . Each

Carriers, Parcel and Delivery

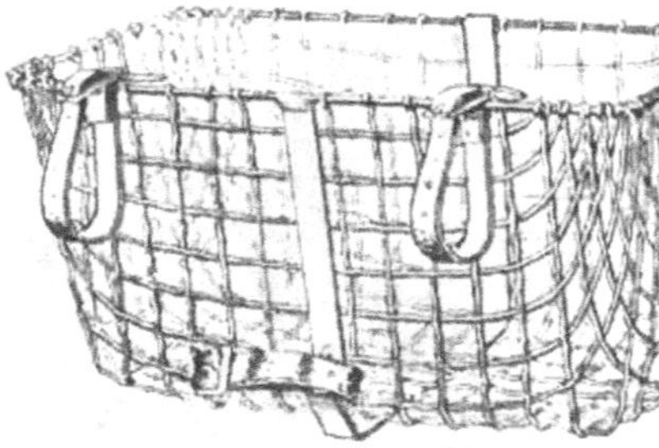

(Made in Canada)

No. 400

The Androck Bicycle Basket.

Just the thing for delivering small packages. Size 15½" x 8" x 7½".

Fastens securely to the Bicycle with strong leather straps

No. 400 . . . . Each

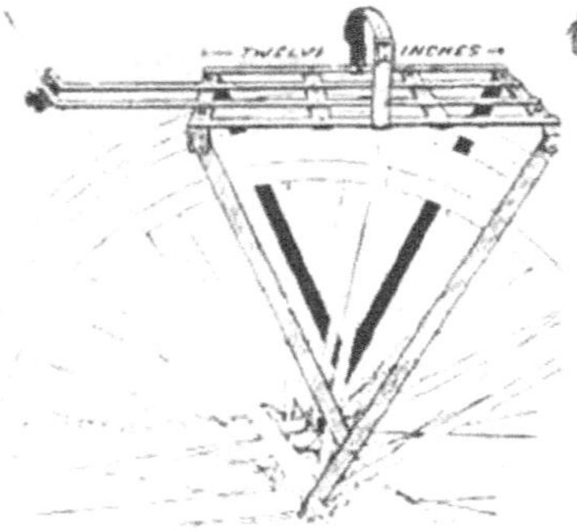

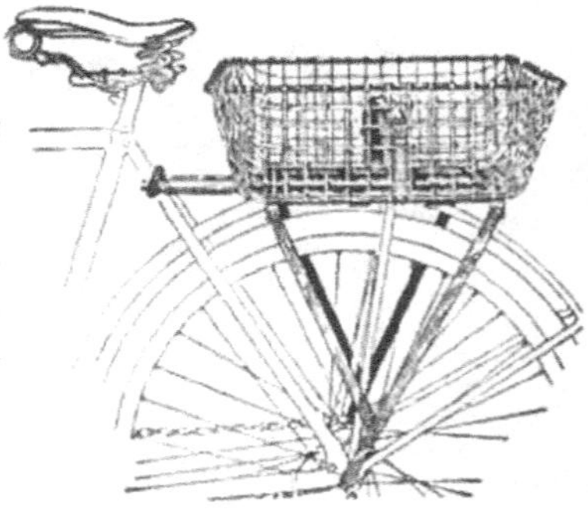

No. 403 *(Made in Canada)* **No. 404**

Androck Luggage Carrier.

Large enough to carry all one wants to. Size of platform, 9" x 12". Length of carrier over all, 18". Supporting legs made of No. 12 steel, ¾" wide. Finest black enamel finish. Size of basket, 18" x 13" x 6".

No. 403—Carrier only (6 in package) . . . . Each

No. 404—Carrier and Basket . . . . Each

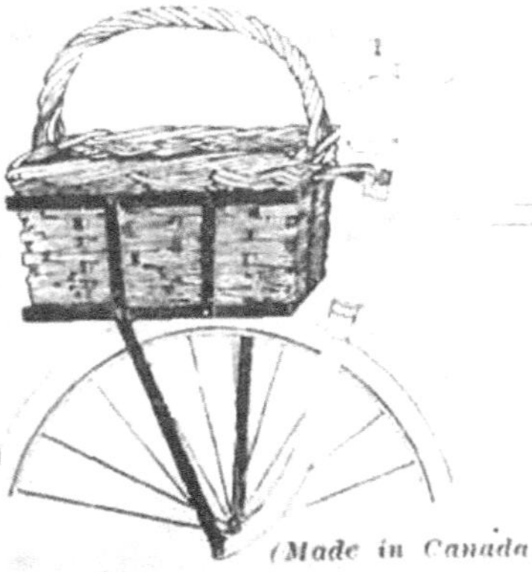

(Made in Canada)

"C.C.M." Tradesmen's.

When fitted with basket makes an ideal parcel carrier. Made of heavy gauge sheet steel, directly supported from front axle. Balances load perfectly, so that bicycle steers easily. Dimensions at top 14" x 20"; at bottom, 12" x 18", 6" deep.

Improved adjustable clamp, adaptable for 20, 22 and 24" frames

No. 529—Carrier without Basket . . . . Each

No. 523—Wicker Basket for above . . . . Each

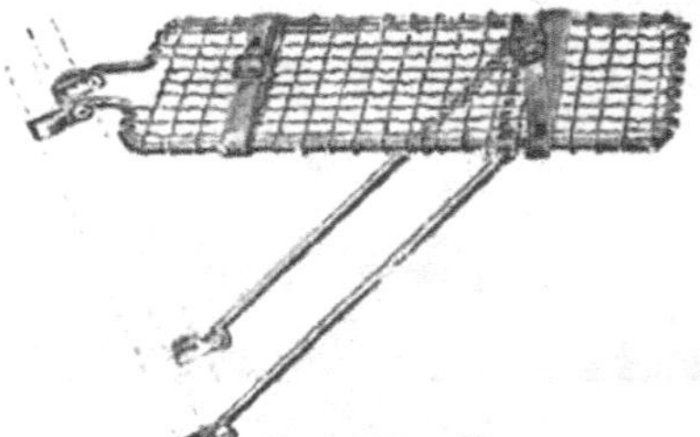

No. 407. Rear.

REAR.

No. 407—Size 14½" x 5½" Split Leather Straps. . Each (12 in package.)

Carriers (Continued)

No. 409 Rear.

No. 409 (6B)—An exceptionally strong, well-made, nickel-plated Carrier, size 12" x 5". Pressed steel top, two leather straps

No. 409 — For Rigid Frame Bicycles. Each

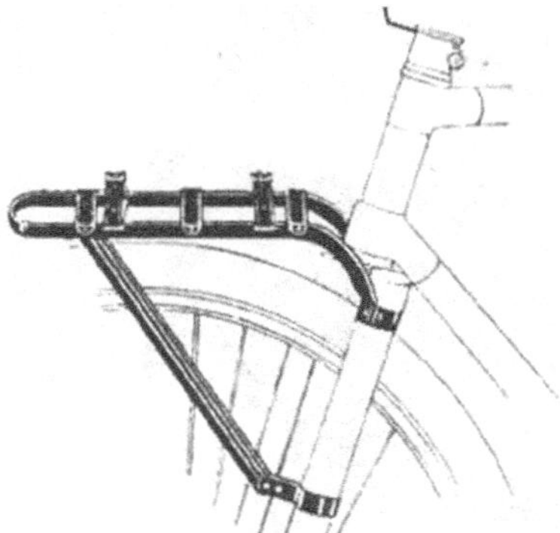

No. 522.

(Made in Our Own Factory)

"C.C.M." Military.

As supplied to the Canadian Army. Unusually strong. Black enamel, with straps. Size 13" x 5¾".

No. 522. ....Each

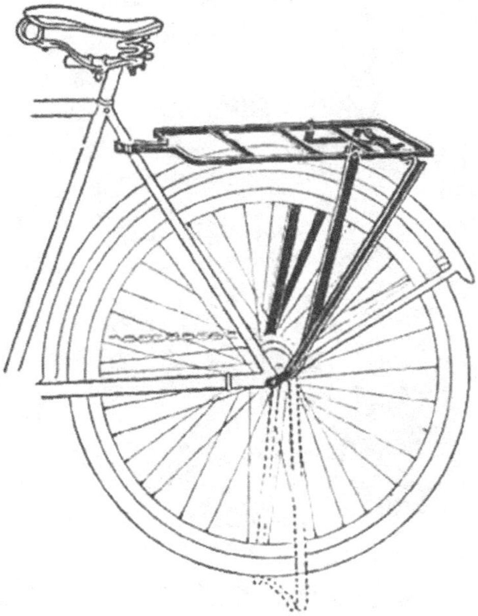

No. 412. *(Made in Canada)*

C.C.M. Combination Rear Carrier and Stand.

The size of both Stand and Carrier is exceptionally generous. The Carrier is both wider and longer than all others. The Stand is about 2" wider. The bottom of the Stand curves upward toward the centre, the sides forming feet, enabling Stand to rest steadily on any surface. The finish is unusually brilliant and durable.

No. 412—C.C.M. Combination Rear Carrier and Stand. Width of feet of stand, 11". .... Each

No. 571—Grey enamel ..... Each

No. 572—Maroon enamel .... Each

No. 563—C.C.M. Rear Carrier (less Stand). Size 7" x 14½". .... Each

Cements

"Comfort" Wood Rim Cement

(Made in Canada)

This Cement is very thick and heavy, and leaves a deep body on the rim, which does not become hard or brittle. Supplied as follows, Imperial measure:

No. 449—2 oz. bottle, with swab (1 doz. in box). Doz.

No. 450—¼ pint cans ... Per doz.

No. 451—1 pint cans ... Per doz.

"Comfort" Wood Rim "Shellac" Cement

(Made in Canada)

For cementing tires and grips. Made from a special formula, so treated as to prevent drying too quickly. Easy to handle, and keeps rims clean. Supplied as follows, Imperial measure:

No. 452—2 oz. bottles with swab (1 dozen in box). Doz.

No. 453—1 pint cans ....Each

Cement, "Titewad"

The Rubber Putty. Permanently repairs rubber tires, tubes and other rubber articles without vulcanizing.

In screw top tin box, containing two small tins, one of putty and one of cement, with full directions.

No. 566—Large size .....Each

No. 567—Small size .....Each

No. 573—25c. Retail Size for Bicycles ....Per doz.

Hard Tire Cement

Old English Brand. Contains an unusually high percentage of rubber and gutta. Supplied in one-pound packages.

No. 456 .... Per lb.

Cements (Continued)

(Made in Canada)

Cement, C.C.M. XX Patching and Fabric

(Imperial Measure.)

No. 431—¾" x 4" collapsible tubes (2 dozen in box). .... Per gross
No. 432—1" x 4" collapsible tubes (1 dozen in box). .... Per gross
No. 433—¼ pint cans. Doz
No. 434—½ pint cans. Doz
No. 435—1 pint cans. Each
No. 436—1 quart cans. Each
No. 437—½ gal. cans. Each
No. 438—1 gal. cans. Each

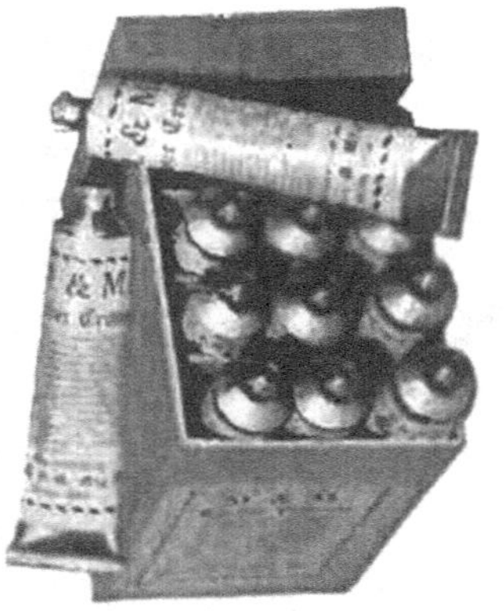

Cement M. & M. Patching

An unusually high grade patching cement.

No. 2335—¾" x 4" collapsible tubes (1 dozen in box). .... Per gross
No. 2336—1" x 4" collapsible tubes (1 dozen in box). .... Per gross
No. 2330—¼ pint. Per doz
No. 2331—½ pint. Per doz
No. 2332—1 pint. Per doz
No. 2333—1 quart. Per doz

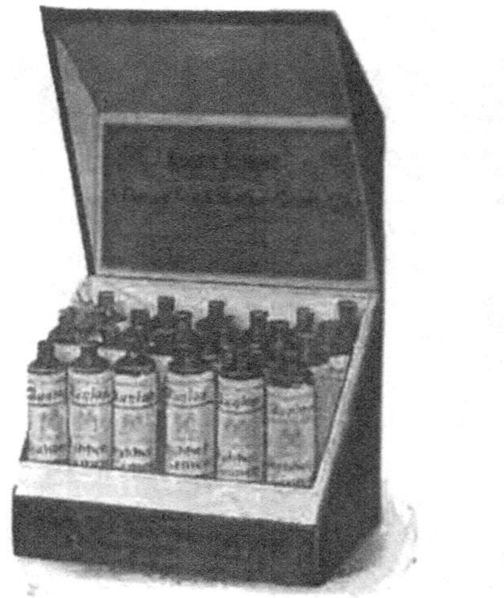

(Made in Canada)

Cement Dunlop Patching

Supplied as follows. Imperial measure:

No. 423—¾" x 4" collapsible tubes (2 dozen in box). .... Per gross
No. 424—1" x 4" collapsible tubes (1 dozen in box). .... Per gross
No. 425—¼ pint cans. Doz.
No. 426—½ pint cans. Doz.
No. 427—1 pint cans. Doz.
No. 428—1 quart cans. Each
No. 429—½ gal. cans. Each
No. 430—1 gal. cans. Each

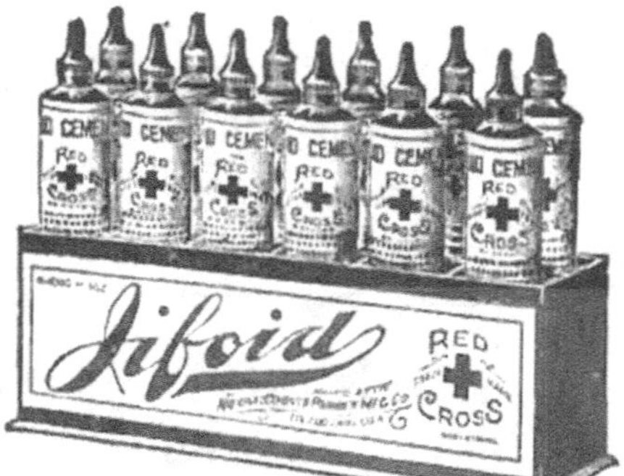

Cement, Jifoid Plugging

"Jifoid" repairs single tube tires in two minutes. By simply injecting "Jifoid" in the puncture and allowing it to harden a perfect and permanent plastic plug is formed.

No. 457—¾" x 4" tubes, with needles (12 in box). Per doz

Dunlop Bicycle Cement Repair Outfit

Includes sandpaper, also French chalk, four round patches, strip of best quality patching rubber, piece of proof cover canvas to patch serious cuts through outer cover. Tube of high-grade rubber cement. Put up in neat design of box, which will fit conveniently in the tool bag.

No. 460 .... Price per dozen

(Made in Canada)

Carbide, C.C.M.

(Made in Canada)

C.C.M. Carbide is put up in handsomely lithographed cans—Blue and Orange—making a very striking article for window display.

No. 413—½-lb. cans. (3 doz. in box) ....Per dozen

No. 414—1-lb. cans. (2 doz. in box) ....Per dozen

No. 415—2-lb. cans. (2 doz. in box) ....Per dozen

No. 416—5-lb. cans. (1 doz. in box) ....Per dozen

No. 417—100-lb. drums.....Each

Auto Carbide **No. 2030—5-lb. cans.** .... Per dozen

Auto Carbide **No. 2028—100-lb.** drums ....Each

Note.—The above cans are full weight.

Chains

Chains—Diamond. (Packed in cartons.)

No. 588—½" pitch x ⅛" wide, 112 links ....Each
No. 589—1" pitch x 3/16" wide, 56 links ....Each
No. 590—½" pitch x 3/16" wide, 112 links ....Each
No. 591—1" pitch x ¼" wide, 56 links ....Each

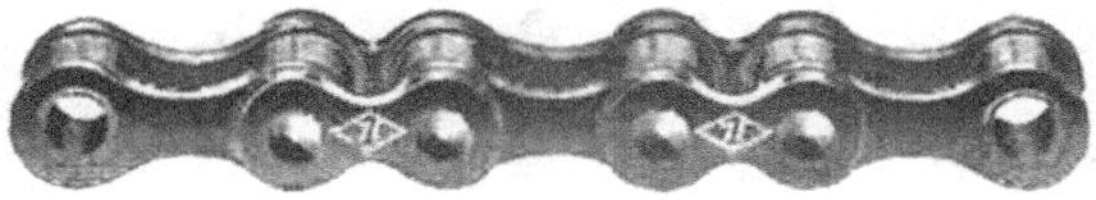

Chain Repair Links

CHAIN REPAIR LINK. "Diamond." For Block Chains only. Simple, effective quick method of adding a link or repairing break in chain.

2 links in package. 1 dozen packages in carton.

No. 480—For 3/16" block chain. Per carton of 1 dozen packages ....

No. 480A—For ¼" block chains. Per carton of 1 dozen packages ....

Diamond "Snap-On" Repair Link

For Bicycle **Roller** Chains. Simple and quick; no riveting.

No. 592 (8·75)—For ½" x ⅛" roller chain ... Per dozen
No. 593 (6·71)—For 1" x 3/16" roller chain .. Per dozen
No. 594 (9·75)—For ½" x 3/16" roller chain .. Per dozen
No. 595 (7·71)—For 1" x ¼" roller chain ...... Per dozen

Chains, Duckworth Motorcycle

25 feet packed in cardboard carton. Sold by the foot.

No. 550—⅜" wide ⅝" pitch .... Per foot
No. 576—¼" wide ⅝" pitch ....Per foot

Chain Bolts and Nuts

(English.)

No. 475—Bolt and Nut for ⅛" Chain, Atco or Maple Leaf ....Dozen
No. 476—Bolt and Nut for 3/16" Chain, Atco or Maple Leaf ....Dozen
No. 477—Bolt and Nut for ⅛" Chain, Brampton ..Dozen
No. 478 Bolt and Nut for 3/16" Chain, Brampton. Dozen
No. 479—Bolt and Nut for ¼" Chain, Brampton ..Dozen

Note.—We do not now stock Atco, Brampton or Maple Leaf Chains complete.

Chain Parts, Roller

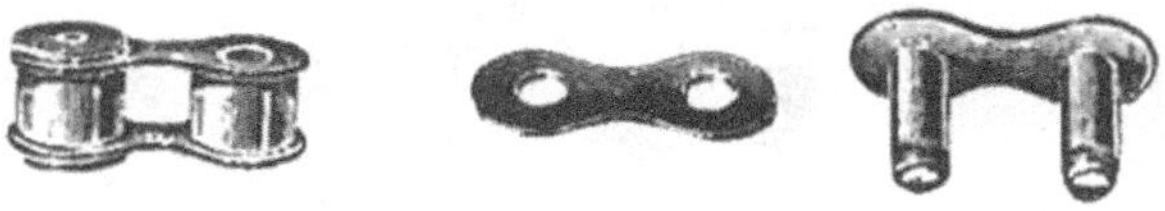

(English.)

Ordinary Repair Link consists of two Rollers, two Bushings, two Inside Plates, one Outside Plate with rivets assembled, one Outside Plate only.

No. 472—For ⅛" x ½" Chains for Atco or Brampton ....Per doz.

No. 473—For 3/16" x 1" Chains for Atco or Brampton ....Per doz.

No. 474—For ¼" x 1" Chains for Atco or Brampton ....Per doz.

Straight Connecting Link consists of two Bolts, two Nuts, two Outside Links; suitable for Atco, Maple Leaf, or Brampton Chains.

No. 471—For 3/16" x 1" Chains ......Per doz.
No. 519—For ¼" x 1" Chains ......Per doz.
No. 470—For ⅛" x ½" Chains ......Per doz.

We cannot guarantee to supply ENGLISH CHAIN PARTS under present conditions, although we have a limited stock on hand, and have more on order which we may receive.

Chain Repair Assortment Atco

Put up in 8-compartment tin box (not in carton as illustrated) containing:

12 each **No. 470-471** Straight Connecting Links.
12 each **No. 472-473** Ordinary Repair Links.
12 each ⅛" and 3/16" Bolts and Nuts.
12 each ⅛" and 3/16" Bushings.
12 each ⅛" and 3/16" Rivets.
12 each ⅛" and 3/16" Rollers.
12 each ½" and 1" Inside Plates.
12 each ½" and 1" Outside Plates.
No. 526—Per box ....Each

NOTE.—Above parts are for Atco and Appleby Chains, but many of them are suitable for Brampton and other chains.

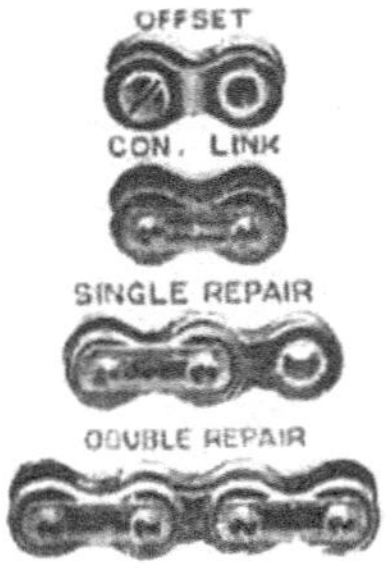

Chain Repair Links

Duckworth Motorcycle

551—Connecting Link ....Each
552—Repair Link, Single ....Each
553—Repair Link, Double ....Each
554—Offset Link ....Each

Chain Parts

Atco Motorcycle

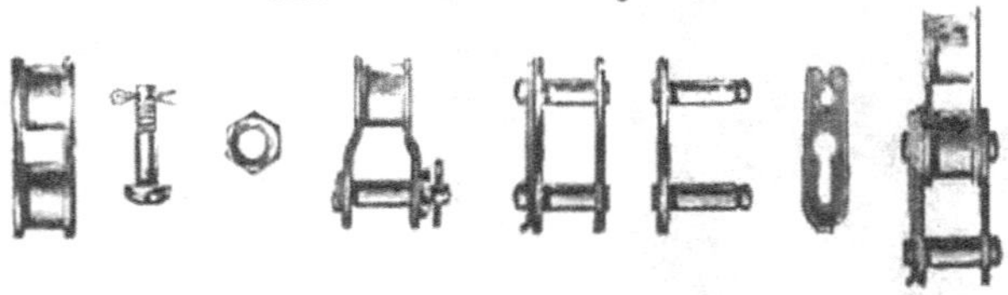

No. 542. No. 534. No. 483. No. 482. No. 536. No. 535. 533.

No. 542—Blocks ....Each
No. 534—Bolts and Nuts ....Each
No. 483—Cranked Link ....Each
No. 482—Connecting Link ....Each
No. 535—Springs ....Dozen
No. 533—Double Link ....Each
No. 536—Rivets ....Dozen
No. 537—Outside Plates ....Dozen
No. 539—Rollers ....Dozen
No. 540 Bushings ....Dozen
No. 541—Inside Plates ....Dozen

Chain Grip—Duckworth

Makes the hard and dirty job of connecting chain very easy, without the necessity of readjusting the wheel. Can be carried in Tool Bag. Black enamelled. One in box.

No. 564 ....Price, each

Chain Tools

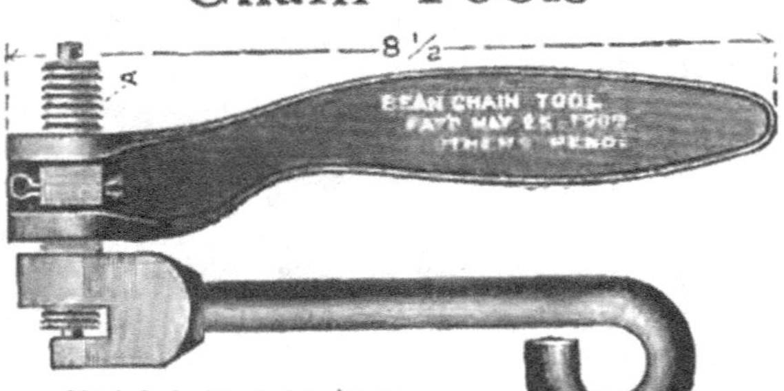

Model A-22. 6 in. long.
Model A-33. 8½ in. long.

Chain Tool, Peerless Motorcycle

"The tool with the punch." Removes any rivet with one grip of the handle. Correct lever action, almost instantaneous and very easily operated.

No. 570 ....Each

Chain Tool, Bean Motorcycle

The handle support has a threaded, headless bolt extended to each side, making it reversible, and giving it double the wear of any other make. (Packed one in a carton.) Complete instructions with each tool.

No. 543—(A-33)—For shop use ....Each
No. 544—(A-22)—Tool kit size ....Each

Clips, Trouser

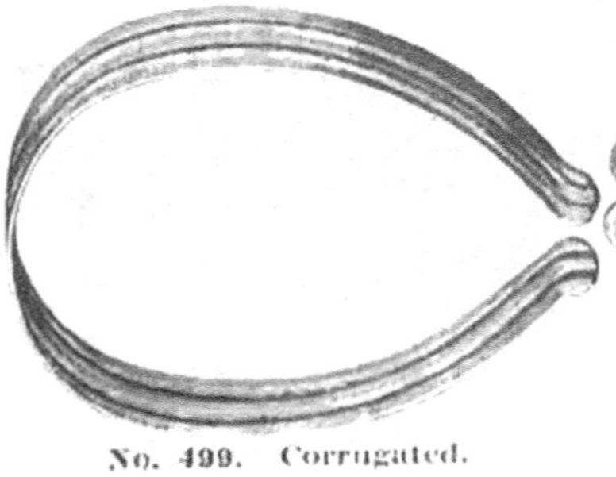

No. 499. Corrugated.

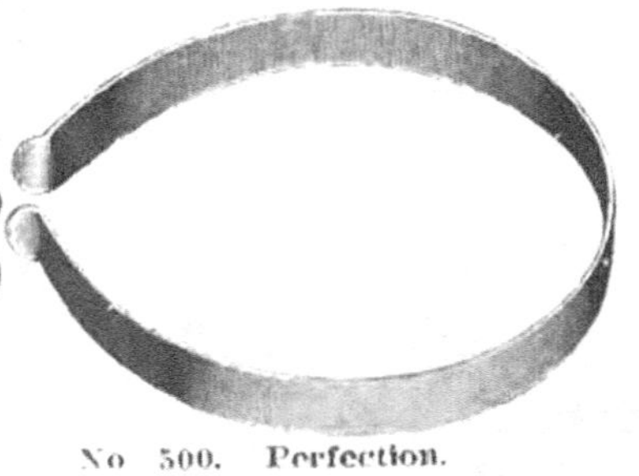

No. 500. Perfection.

Manufactured from the best quality of spring steel and oil tempered with an enamel finish.

No. 499—Per gross pairs (3 doz. pr. in box) ....

No. 500—Wide Perfection Guards. Per gross pairs (3 doz. pr. in box) ....

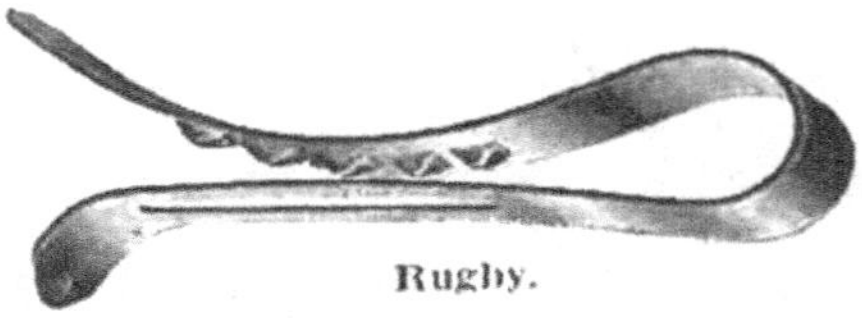

Rugby.

The Rugby Clip—We have no hesitation in saying this is one of the best made. Grips the trousers hard and stays on.

Furnished in either blue lacquered or nickel finish.

No. 497—Nickelled, per gross pairs (3 doz. pairs in box)...

No. 498—Lacquered, per gross pairs (3 doz. pairs in box)...

Clip and Trouser Guard, Eclipse

Pantasote guard, with pant clip inserted. Removed with slight pull at top. Can be comfortably concealed beneath trouser leg when not in actual use.

No. 530—Motorcycle size. Per dozen pairs (6 pr. in carton). Dozen pairs ....

No. 527—Bicycle size. Per dozen pairs (6 pr. in carton) ....Doz. prs.

Clips, Toe

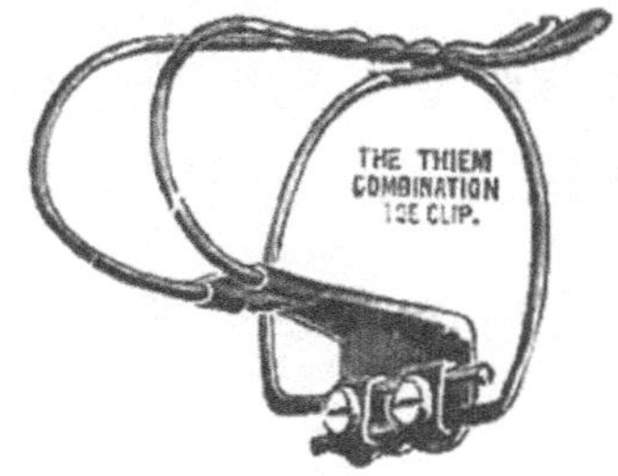

The "Star" Combination.

Made of sheet steel and crucible steel wire, nickel plated and polished.

No 495—Per dozen pairs (1 doz. pr. in box) ..Doz. pr.

Cotter Pins, Crank

Assorted

Furnished complete with nut, well nickeled and polished. Furnished in following sizes: 5/16", 11/32", 3/8", 13/32".

No. 1415—Price (1 gross in box) ....Per dozen

Cyclometers, Veeder Bicycle

No. 511—10,000 mile.

Veeder Bicycle Cyclometers come in two styles. **The 10,000 Mile** is dust proof, water proof, absolutely correct, small and light, every part being as accurately made as the parts of a watch.

The Trip style is practically two cyclometers side by side. One records the total mileage, the other records the trip.

Each in small carton (12 in large counter display carton).

No. 514—10,000 Mile ....Each

No. 515—Trip ....Each

No. 515—Trip.

CYCLOMETER BRACKET

Veeder Bicycle

No. 516 ....Each

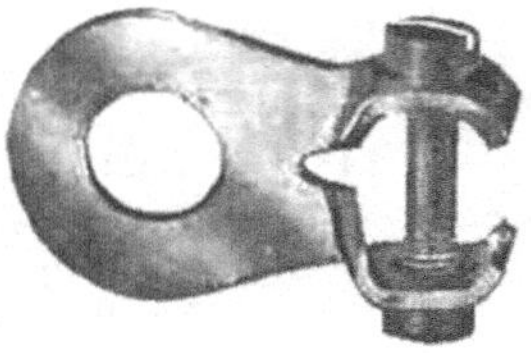

CYCLOMETER STRIKER

Veeder Bicycle

No. 517 ....Each

Cyclometers

Veeder Motorcycle

No. 511—Trip.

The Trip style is practically two cyclometers side by side. One registers the total mileage, the other records the trip. Packed each in a small carton, with complete instructions. (12 small cartons in large carton.)

No. 511—For 26" wheel, including Bracket and Striker (12 in box) ....Each

No. 520—For 28" wheel, including Bracket and Striker (12 in box) ....Each

No. 518—Special Harley-Davidson Model for 1916, 1917 and 1918 Motorcycle, complete with bracket and Striker (12 in box) ....Each

NOTE—When ordering, state serial number of Bracket required.

No. 518.

Cyclometer Brackets, Motorcycle

No. 513—H-10 Bracket.

No. 546 — (H-7) — For Indian, Pope and Excelsior machines ......Each

No. 513 — (H-10) — For all other makes except Harley-Davidson. Each

Note. — Harley-Davidson Motorcycles require special brackets, which are always sold complete with H.-D. Cyclometers.

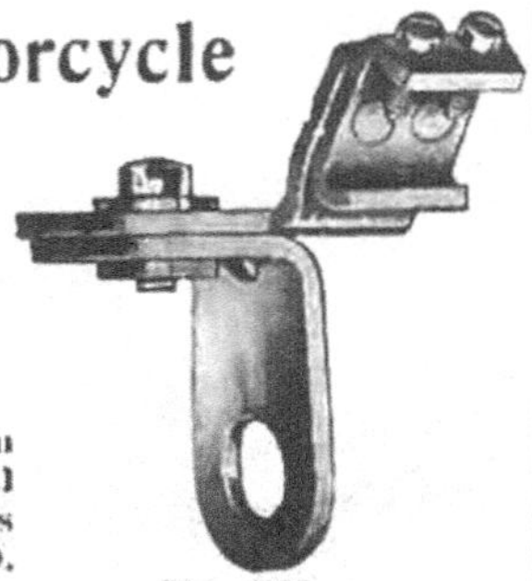

No. 546.
H-7 Bracket.

Cyclometer Striker

Motorcycle

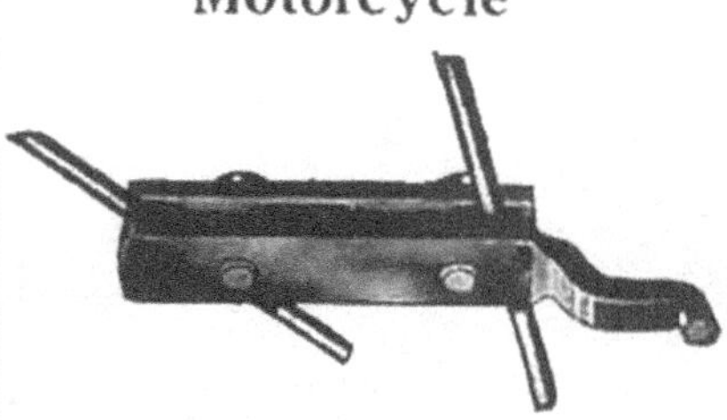

No. 512 ....Price, each

Cones
Adjustable

No. 502—⅛", 24 Thread ....Each
No. 503—[illegible]" 24 Thread ....Each

Crank Repair Tips

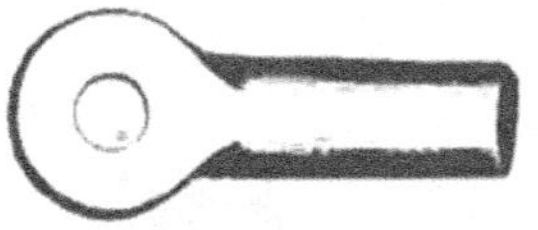

For repairing cranks broken at the pedal end. Drop forgings.

Threaded ½ x 20, in pairs, right and left.

No. 509R—Right ....Each
No. 509L—Left ....Each

Cushion for Passenger Carrying (Motorcycle)

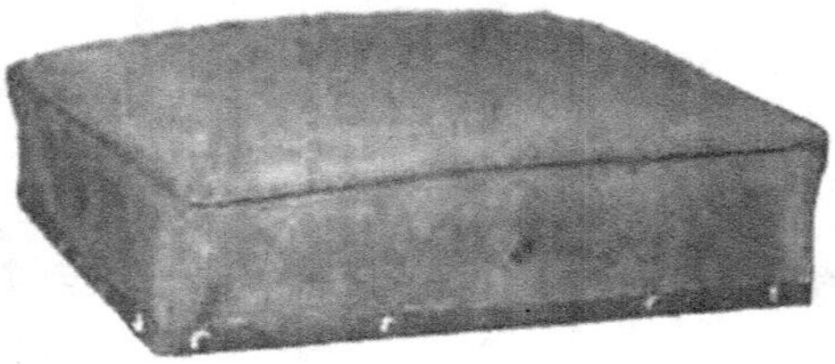

This is a spring cushion, with soft, stuffed cover and board bottom. Finished in leatherette with folded corners and full bound base. Furnished in red and brown. Size 14" x 8¾" x 4½". Fitted with four attaching brackets.

No. 411 ....Each

Chains, Tire
Weed Motorcycle

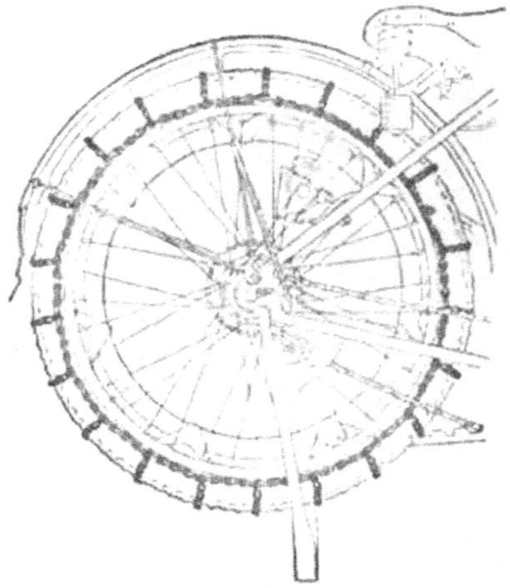

Made for service. Are thoroughly practical and safe. Easily attached. Take up little room in tool bag, and can be carried in pocket. Packed one in a canvas bag.

No. 485—26 x 2½ ....Each
No. 486—28 x 2½ ....Each
No. 487—28 x 2¾ ....Each
No. 488—29 x 2¾ ....Each
No. 489—28 x 3 ....Each

Cross Chains.

No. 490—2½" ....Each
No. 491—2¾" ....Each
No. 492—3" ....Each

Clamp, Chalfant

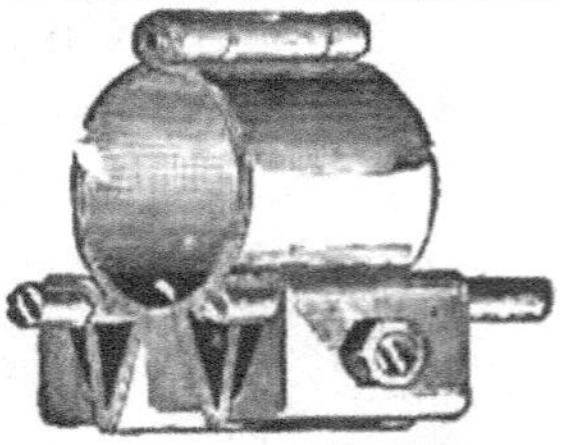

A Clamp for attaching to top bar of frame for purpose of lowering saddle by bringing it closer to the frame than the seat post would allow. No seat post is necessary when this clamp is used.

No. 574—1" size ....Each
No. 575—1⅛" size ....Each

Drivers, Screw

Made from smooth, flat steel with rounded corners. Blade is hardened and point ground. Nicely nickeled.

No. 18043 Per dozen (50 in box) ....

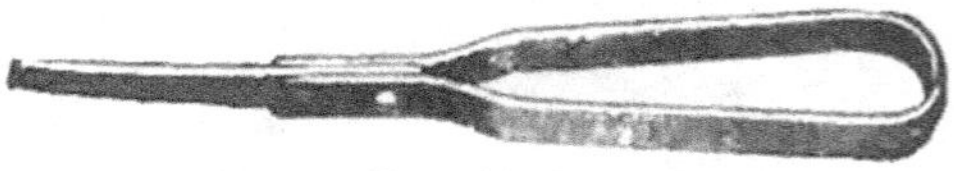

No. 18043.

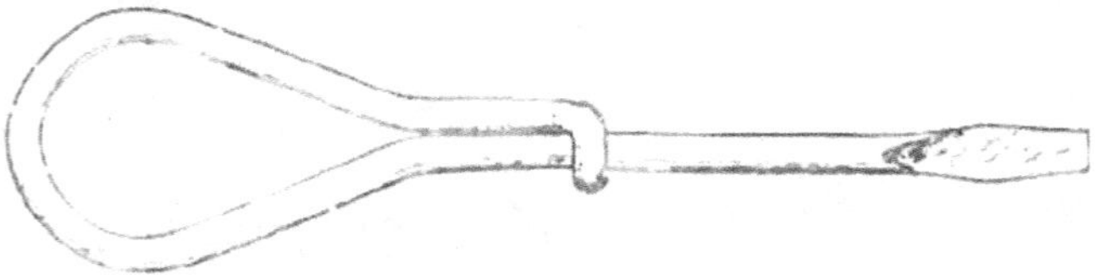

No. 655.

Wire.

This driver is made from the best quality of steel wire, carefully tempered.

No. 655 Per dozen (1 gross in box) ......

No. 18235 Bridgeport, wood handle, 6" hexagon blade (12 in box) .... Per dozen

No. 18235.

Yankee Ratchet, 5 in. Blade.

Material and workmanship are of superior quality in every detail. Strong, durable, handsomely finished; thoroughly tested. Nickel plated. Right, Left and Rigid.

No. 18044—Price (6 in box) ....Each

No. 18044. Right, Left and Rigid.

Drivers, Screw (Continued)

Yankee Ratchet, 1¼ Inch Blade.

Made for special use of gunsmiths, fitters, electricians and mechanics, requiring a strong, substantial screw driver with a short stub blade. It is right or left hand, and rigid.

No. 18045—Each (6 in box) ....

No. 18045.

Spiral Ratchet Screw Driver, No. 657.

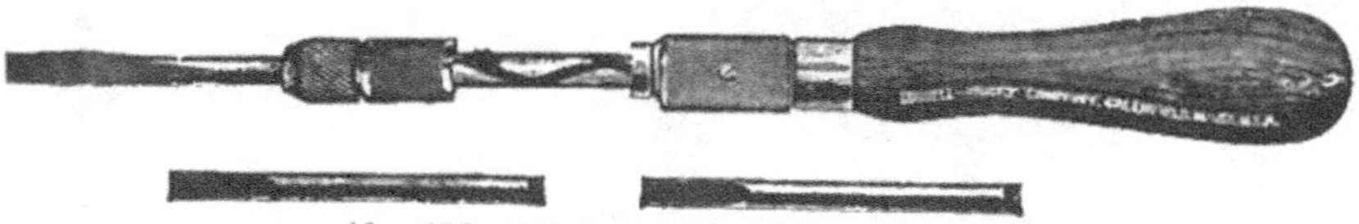

No. 657. Right and Left Hand and Rigid.

Right and left hand and rigid. Can be set rigid at any part of its length.

No. 657—Price (each in carton) ....Each

Spoke Nipple Screw Driver, No. 658.

It is beautifully finished, and is the very best tool for bicycle spoke nipples. Centre of blade is hollow to allow for spoke. Handle is not fluted, as shown in cut.

No. 658—Per dozen ....

No. 658.

No. 659.

Vlchek Motor Cycle Screw Driver

Square shanked for turning tight screws with aid of wrench. Length of blade 5". Length over all, 10½". (6 in box).

No. 659—Price ....Per dozen

Die Sets (See page 61)

C.C.M.

Supplied in the following colors: Black, "Indian" Motorcycle Red, Robin's Egg Blue, Maroon, Tribune Blue, Bullfrog Green, Golden Brown, White, Excelsior Gray, Excelsior Service Green, Harley-Davidson Service Green, Harley-Davidson Gray, French Gray.

No. 703—¼ pint tins (wine measure), **Black** ....Per dozen

No. 704—¼ pint tins (wine measure), **Colors** ....Per dozen

No. 714—¼ pint tins, (wine measure), **White** ....Per dozen

No. 705—1 pint tins (wine measure), **Black** ....Each

Enamels, Air Drying

(Made in Canada)

Club Black

Special Hard Drying

(Two Sizes.)

This enamel is very convenient for repair shop use, as it dries hard in fifteen or twenty minutes.

No. 701—Club—Small size (2 gross in box). Per dozen

No. 702—Club—Large size (1 gross in box). Per dozen

Club Enamel.

Enamel
Lowe's

(Made in Canada)

A high grade, air-drying enamel, which dries with a beautiful brilliant lustre.

No. 713—¼ pint cans, **Imperial** measure, black only ....Dozen

Enamel, Baking

No. 710 Black (Imperial measure), 1 pint cans ....Per dozen

(Made in Canada)

Aluminum Paint

Dealers will find this Paint an all-round seller for decorative purposes. A beautiful and untarnishable finish for frames, forks, wood rims, guards and nickel work. Has the appearance of frosted silver. Supplied as follows:

No. 711—Aluminum, ¼ pint ......Per dozen

No. 712—Aluminum, ½ pint ......Per dozen

Frames and Forks, Model T

(Made in Our Own Factory)

T Frame, Forks and Fittings, as above (Black) ....
This frame used in all C.C.M. Bicycles. We manufacture one frame only.
Extra for Colored Enamel ....
Extra for Colored Sunburst Head ....
Extra for Striping ....

"Hercules" Repair Forks

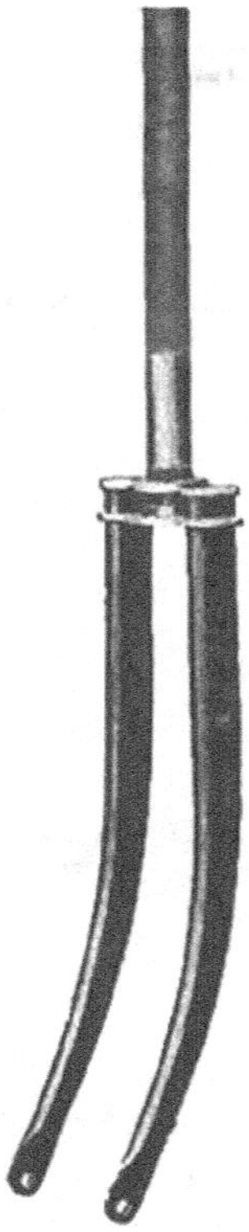

(Made in Our Own Factory)

Fork Stems only

(Made in Our Own Factory)

No. 756—24 Thread for Canadian Bicycles. ....Price, each
No. 757—30 Thread for Canadian Bicycles. ....Price, each
No. 758—26 Thread for English Bicycles. ....Price, each

"Hercules" Repair Forks

We supply these as per following specifications:
Fork sides and crown enamelled black; 10" stem.
No. 749—Double Plate Crown, 24 thread, enamelled ....Each
No. 755—Double Plate Crown, 26 thread, enamelled ....Each
No. 750—Double Plate Crown, 30 thread, enamelled ....Each
No. 761—Double Plate Crown, not threaded, enamelled ....Each

NOTE.—Above are now threaded for 26" frames, and can be fitted to 20", 22" and 24" frames by cutting to the necessary length.

Fork Stem Set, Derby

(Made in Canada)

Die and Guide—2¾" diameter, Cutting Thread 1" x 24", with 26" stock. Packed in wooden box; Set Screw packed in end of Die Holder.
No. 2115—Each ....
Extra Dies, for Cutting Threads 1" x 24, 26, 28 or 30 stock ....Each

Emery Cloth

We stock in sheets 9" x 11" B. & A. Twilled Black Emery Cloth, as below:

No. 4003—(Fine) (1 quire pkg.). Per doz.

No. 700—(Medium) (1 quire pkg.). Per doz.

No. 4004—(Coarse) (1 quire pkg.). Per doz.

Fork Repair Tips

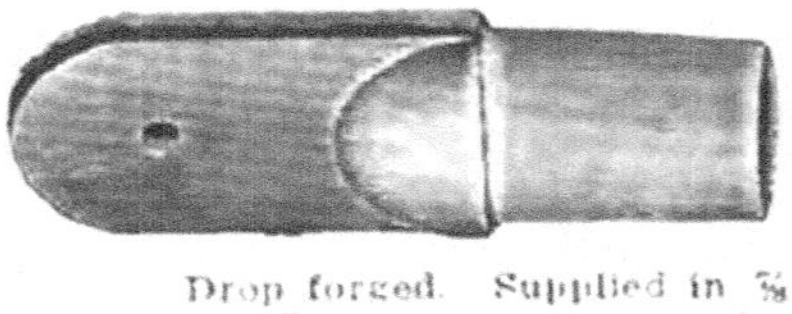

Drop forged. Supplied in ⅞" and 1".
No. 2120—⅞" ....Per dozen pair
No. 2121—1" ....Per dozen pair

DEALERS' NAMEPLATES.

If you require a special nameplate, send us your sketch with particulars, and we will submit a drawing and definite quotation. **Regular type nameplates** finished in one color, a choice of nickel-plated, oxidized silver or gold plated.

25 Nameplates ..
50 Nameplates ..
100 Nameplates ..
200 Nameplates ..

Extra for two-colored plate. .Per 100
Extra for enamelled name ....Per 100

Frame Parts for Hygienic Cushion Frames

No. A.—Rear Upper Hygienic Fork Assembly, stripped. .... Each
No. B.—Dust Cap and Head. .... Each
No. C.—Plunger Tube. Each
No. D.—Cartridge .... Each
No. E.—Brass Bushing. Each
No. F.—Spring ....Each
No. G.—Steel Washer. Each
No. H.—Leather Washer. .... Each

(There is a leather plug fitted in the rear upper fork assembly which is used as a cushion for the cartridge, eliminating any rattling noise.)

Note.—When ordering Spring (give number), state whether for 22" or 24" frame.

No. 1, for riders weighing 100 lbs. to 125 lbs.

No. 2, for riders weighing 125 lbs. to 150 lbs.

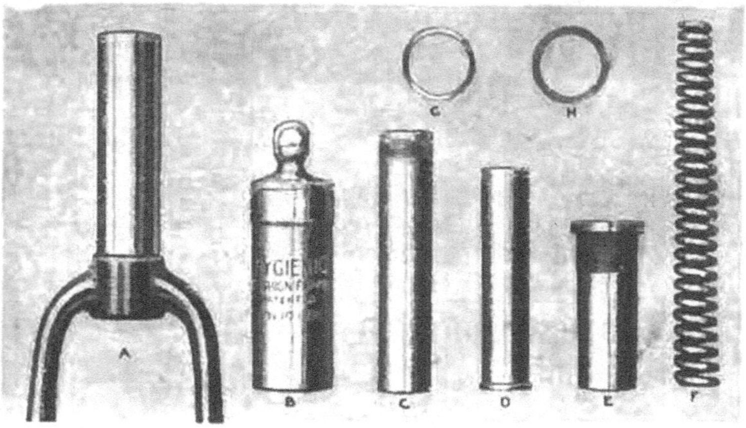

(Made in Our Own Factory)

No. 3, for riders weighing 150 lbs. to 175 lbs.

No. 4, for riders weighing 175 lbs. to 200 lbs.

No. 5, for riders weighing 200 lbs. to 250 lbs.

Much of the benefit from this splendid feature is lost if the spring is not fitted to the rider's weight.

Fluids, Tire

"Neverleak."

Per doz.

No. 751—4 oz. tubes (12 in box)......

Grips, Handlebar

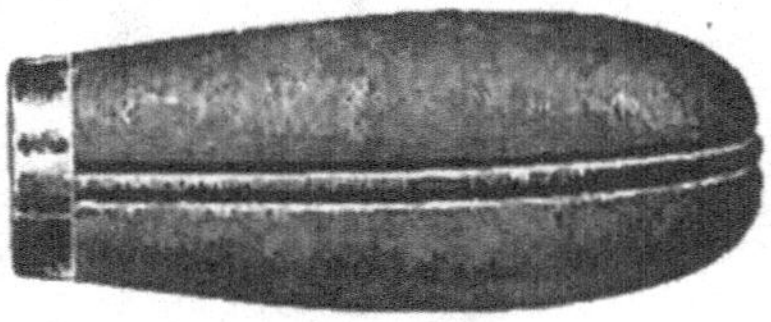

(Made in Canada)

No. 823.

Sewn Leather Grips, Brown. For ⅞" bars.

No. 823—(1 doz. prs. in box). ....Per pair 3.265

No. 822—4.

No 824 (75)—Leather wound, ⅞" nickeled ferrules (1 doz. prs. in box). Doz. prs.

No. 822 (75)—Leather wound, ¾" nickeled ferrules (1 doz. prs. in box). Doz. prs.

No. 825.

No. 825 (68)—Leather wound, wood tips (1 doz. prs. in box) ....Doz. prs.

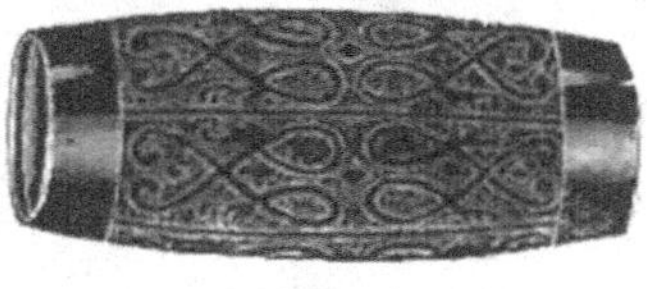

No. 826.

No. 826 (171)—Celluloid Bluemel (English manufacture) (1 doz. prs. in box) .....Dozen pairs

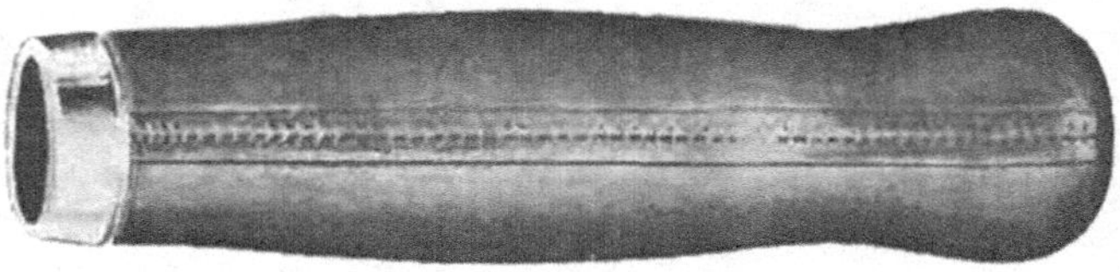

No. 828.

No. 828—Bull Dog. First quality sewn leather grips, ⅞" (1 doz. prs. in box) ....Pair

Grips, Handlebar (Continued)

Liberty

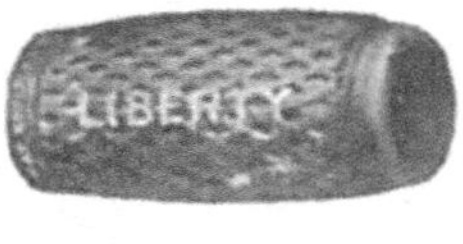

A rubber grip of neat design, as illustrated, square ends, standard size.

No. 888 (157)—(1 doz. prs. in carton) ....Per pair

Corrugated Rubber

This is a moulded grip made of rubber throughout, reinforced with heavy fabric inside, and fitted with strong nickeled ferrule. Length over all 3¾"

No. 874.

No. 874 (94)—(1 doz. prs. in box) ......Per pair

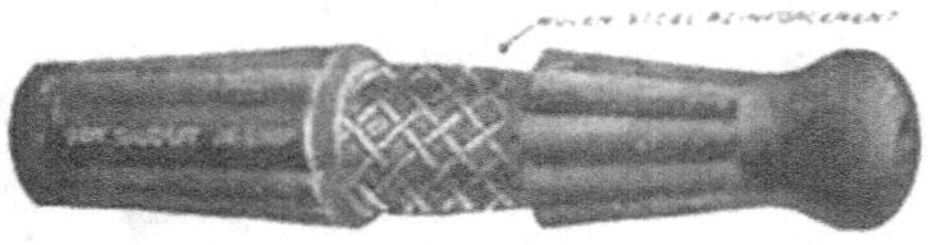

No. 829.

Boy Scout, Jr.

No. 829—Boy Scout, Jr., woven steel reinforced rubber grips, ⅞" (1 pr. in box) ....Pair

Enduro Black Rubber

A well designed and popular priced Rubber Grip.

No. 882—One dozen pairs in a carton .....Per pair

Motorcycle Woven Steel Reinforced

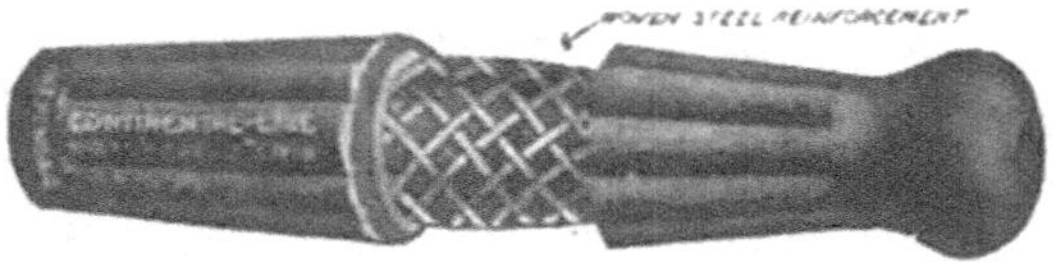

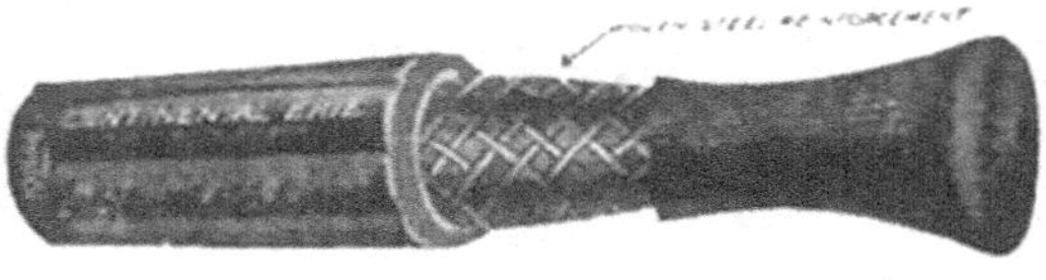

These grips are made with an inner lining of fabric and steel reinforcement. Made in two styles as shown.

Regular Type—No. 830 (200)—Regular, 1" (one pair in box). Pair

No. 831 (200)—Regular, 1⅛" (one pair in box) Pair

Indian Type—No. 864 (100)—For 1" bars. Corrugated. Special for Indian Motorcycle (one pair in box) ....Per pair

Grips, Nipple

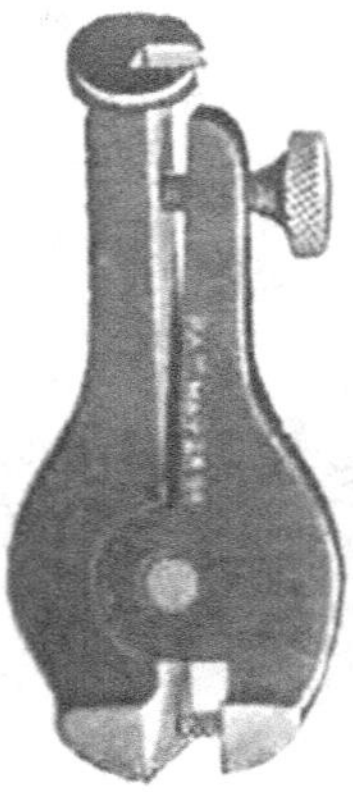

Perfect.

Perfect.

Made in three sizes

No. 832 (1)—Small, for general use (1 doz. in box). ....Each

No. 833 (2)—Large, for repair work (1 doz. in box). ....Each

No. 834 (3)—Auto, for heavy work (1 doz. in box). Each

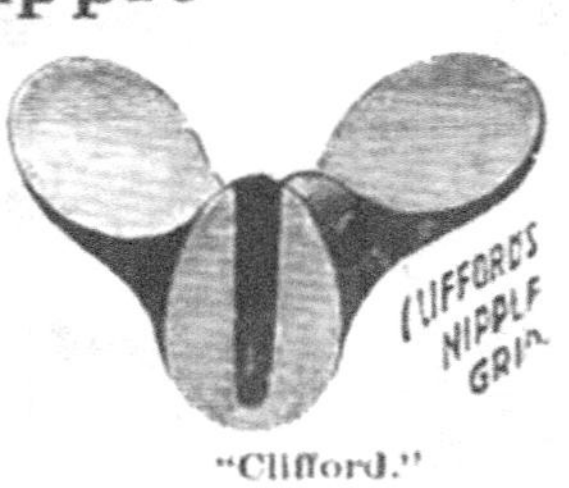

"Clifford."

Nickel Plated.

No. 836—Clifford

.........Per doz.

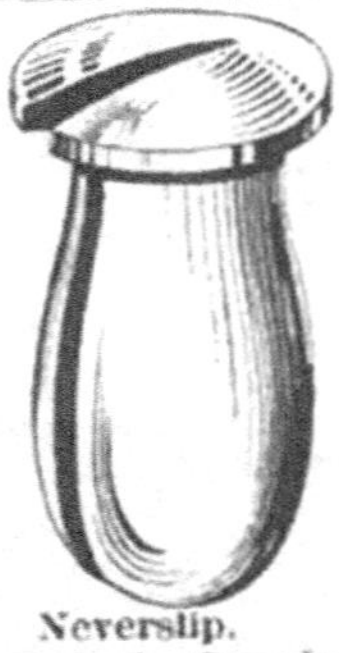

Neverslip.

(Made in Canada)

The harder you pull the tighter it holds. Heavily nickel-plated, and buffed.

No. 835—Per doz. .

Guards

(Made in Our Own Factory)

These guards are all made from selected maple and oak. They are all highly finished. **Eyeletted** unless otherwise ordered.

Guards, Wood Dress and Chain

No. 850—Dress only, plain maple ....Each
No. 851—Dress only, plain oak ....Each
No. 852—Chain only, plain maple ....Each
No. 853—Chain only, plain oak ....Each

Made of celluloid, light yet durable, and a very effective article for keeping mud off your clothes.

Bluemel Side Mud Guards

No. 867—Front and Rear (packed 3 pairs in one box) ....Pair
No. 868—Front only (2 in box) ....Each
No. 869—Rear only (3 in box) ......Each

Guards

Steel, Mud

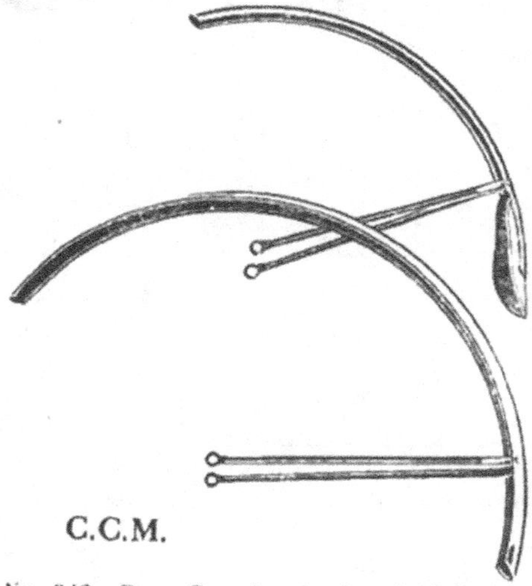

C.C.M.

C.C.M., with flaps. Black enamelled steel guards, equipped with heavy leather oval flaps and steel braces, as supplied on C.C.M. Grade A Bicycles.

No. 840—Front and Rear, with flaps for rigid frame ...Per pair
No. 841—Front and Rear, with flaps, for hygienic frame ...Per pair
No. 842—Front Guard only, with flap.. Each
No. 843—Rear Guard only, for rigid frame ....Each
No. 878—Rear Guard only, for hygienic frame ....Each
No. 886—Oval flaps only ....Per dozen
No. 887—Braces only ....Per dozen
Extra for gold striping on above ....Per pair
Extra for maroon and gold striping on above ......Per pair
Extra for other colors with gold striping ....Per pair

NOTE.—Only black and maroon guards carried in stock; other colors made to special order.

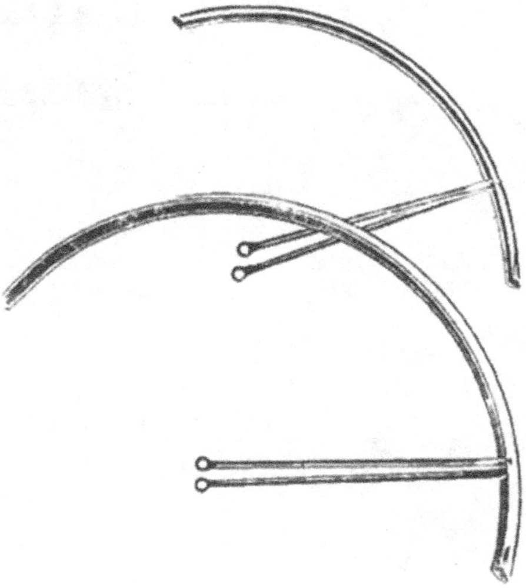

Comfort, Less Flaps

No. 885—Front and Rear, less flap, for rigid frame ....Per pair
No. 884—Front Guard only, less flap, for rigid frame ....Each
No. 843—Rear Guard only, less flap, for rigid frame ....Each
No. 887—Braces only ....Per dozen
Extra for gold stripping on above. Per pair
Extra for maroon and gold striping on above ....Per pair
Extra for other colors with gold striping.. ....Per pair

NOTE.—Only Black and Maroon Guards carried in stock; other colors made to special order.

Tubular Rivets

For C.C.M. Mud Guard Flaps ....Per 100 **$0.60**

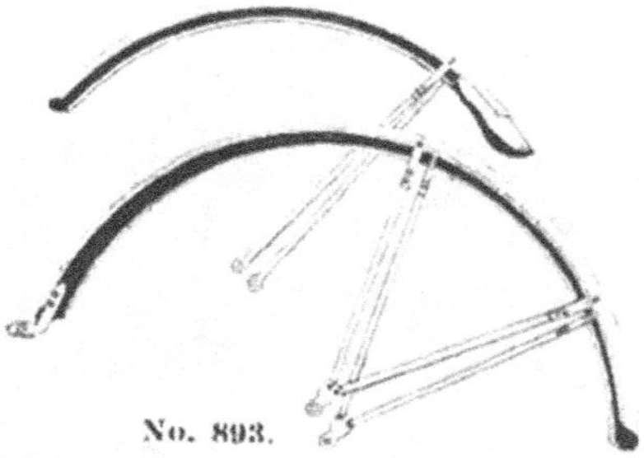

No. 893.

DOUBLE BRACE MUD GUARDS, for Model "W," C.C.M. Motorbike.

No. 893—Front and Rear, with flap, double brace on rear guard. Black ....Per pair
Extra for green and gold stripe, or red with green stripe, the two standard colors of C.C.M. Motorbike ....Per pair

Guards, Steel, Dress and Mud

No. 890—C.C.M. Front and Rear, with flap, eyeletted ....Per pair
(No. 842 Front Guard used.)
No. 891—Comfort Front and Rear, less flap, eyeletted ....Per pair
(No. 884 Front Guard used.)
No. 892—Rear Guard only, eyeletted......Each
No. 842—C.C.M. Front Guard only, with flap ... .Each
No. 884—Comfort Front Guard only, less flap.... ....Each
No. 886—Flaps only ....Per dozen
No. 887—Braces only ....Per dozen
Gold striping and colors extra as for C.C.M. and Comfort Men's Guards.

Guards, Chain, Metal

Metal Chain Guard, with Parts.

No. 838—Complete with fittings ....Each

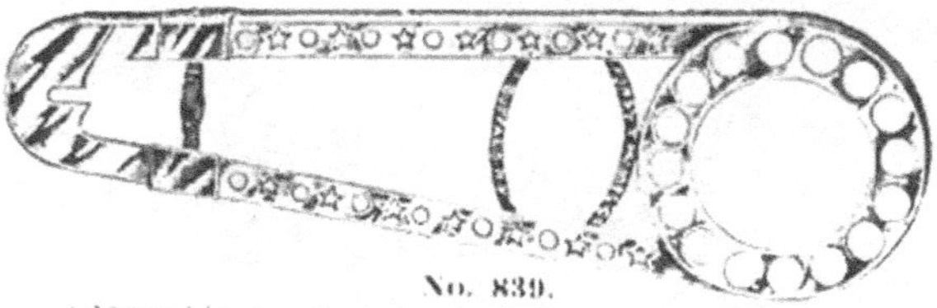

No. 839.

Adjustable to stretch of chain and to difference of length between centres of axles.

No. 839—With attaching parts ....Each

Chain Guards, C.C.M.

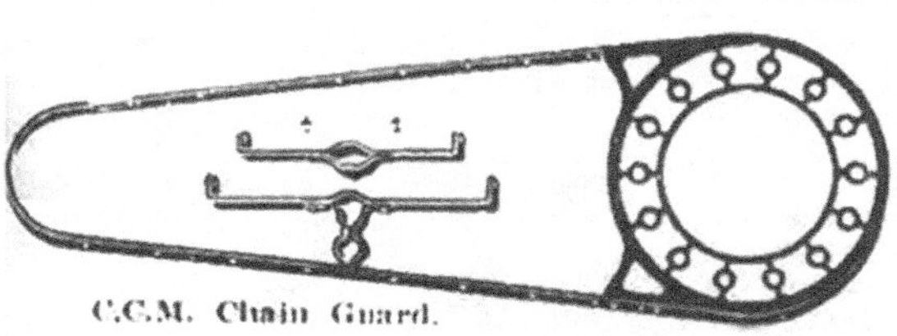

C.C.M. Chain Guard.

As equipped on C.C.M. Ladies' Bicycles.

Metal Disc type, as illustrated, completely enclosing sprocket, and with continuous maple rears, eyeletted ready for lacing.

No. 889—Complete with fittings, full black enamel... ....Each

Only Black and Maroon Chain Guards in stock; other colors made to special order.

Guards, Chain and Dress Guard Parts

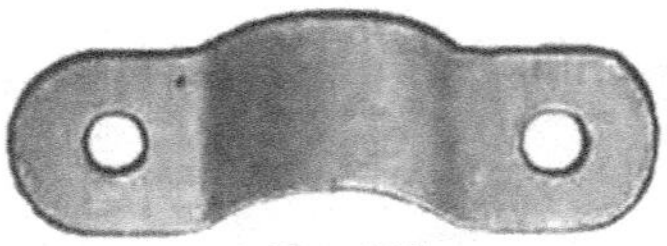

No. 857.

Top Connection, including Bolts and Nuts ....Doz.

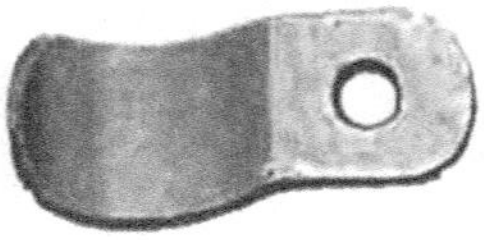

No. 858.

Bottom Connection, including Bolts and Nuts ....Doz

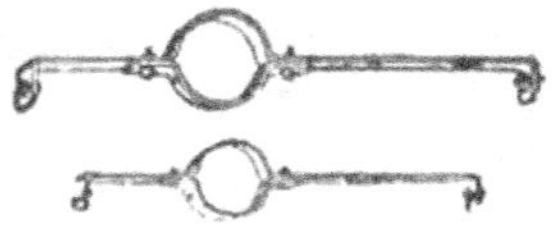

No. 860.

No. 861.

No. 860—Front Connection, nickel plated ....Dozen

No. 861—Rear Connection, nickel plated ....Dozen

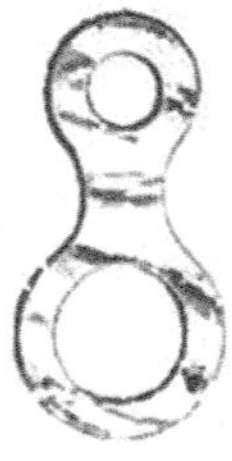

No. 863.

No. 863 — Lacing Loops, nickel plated.

Doz. pr..

No. 859.

No. 859—Front Connection, complete with round head machine screw ...Per doz.

No. 862.

No. 862—Dress Guard Brace, round, nickel plated (25 in package) ....Dozen

Guard Lacing, Silk

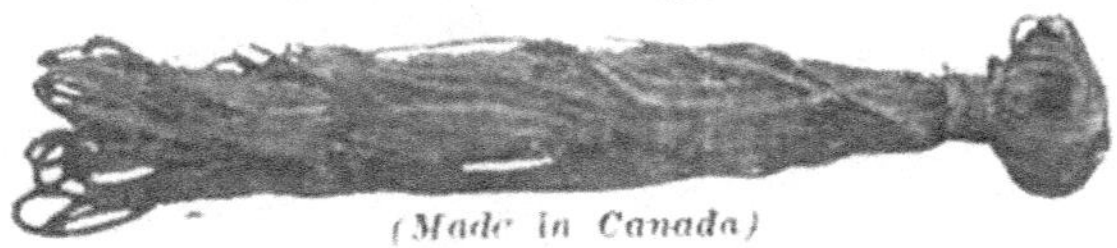

(Made in Canada)

Supplied in 18-yard hanks, in black or in Cleveland brown.

Black silk guard lacing is standard equipment on all 1917 and 1918 C.C.M. Bicycles. Cleveland Brown was standard equipment on 1916 bicycles.

No. 1105—Black ....Per doz.
No. 1106—Cleveland brown ....Per doz.

Guard Lacing, Detachable

Each net is firmly attached to a nickel plated bracket or clip, which is designed to be placed beneath the nut on the rear axle. These nets are waterproof.

No. 1115—Per pair (1 doz. in box, assorted) .....

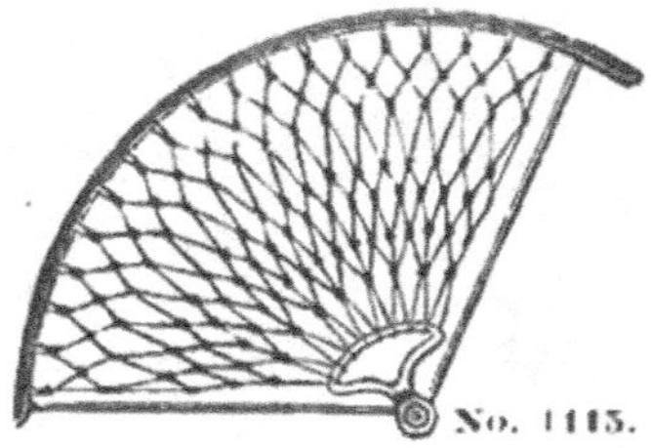

No. 1115.

Gauge, Screw Pitch

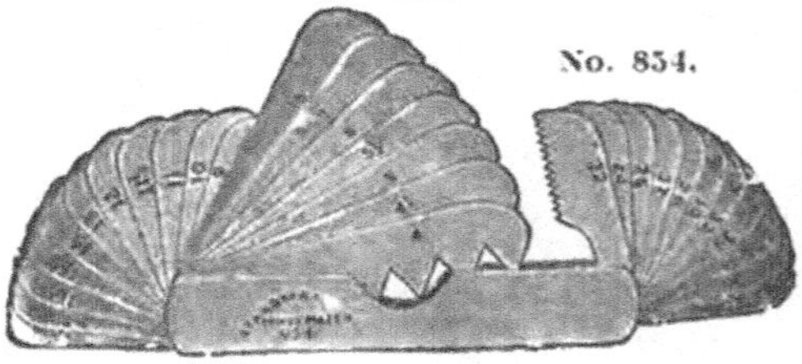

No. 854.

For gauging the pitch of inside and outside threads. All pitches from 4 to 40.

No. 854 -Price ....Each

(Made in Canada)

Holder for Skate Grinding

An adjustable, easily operated Skate Grinding Holder, which is accurate. The bottom plate is fitted with ball bearings, which make it run very easily. Improved design.

No. 881 ....Each

Hack Saw Blades

Flexible Backs

The teeth only are hardened, the back remaining soft, i.e., not tempered. The blade will neither snap nor break. Supplied in 18 teeth to the inch for ordinary work, and 32 teeth for cutting tubing. All blades come in packages of 72 each.

No. 18011— 8", Coarse ....Dozen
No. 18012— 9", Coarse ....Dozen
No. 18013—10", Coarse ....Dozen
No. 18014—11", Coarse ....Dozen
No. 18015 12", Coarse ....Dozen
No. 18006— 8", Fine ....Dozen
No. 18007— 9", Fine ....Dozen
No. 18008—10", Fine ....Dozen
No. 18009—11", Fine ....Dozen
No. 18010—12", Fine ....Dozen

Hack Saw Frames

(Made in Canada)

Lock Extension—Made of best quality steel, handsomely finished and nickeled. Takes blades from 8 to 12 inches, inclusive. One blade supplied.

No. 18020—(2 in box) ....Each

Prices Subject to Change Without Notice.

Handlebars

For illustrations of H.B. Stems separately and for size specifications of H.B. Tops, see following pages.

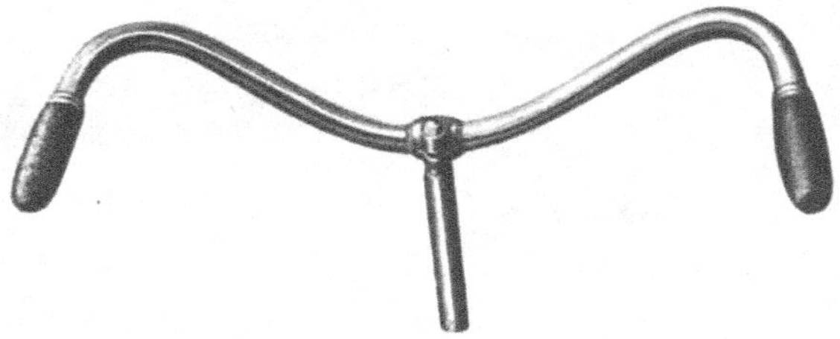

No. 955. *(Made in Our Own Factory)*

No. 955—(5½)—**C.C.M. Regular** H. Bar and Post Complete, less Grips .... Price, each

No. 956—20" Top .... Price, each

No. 957—C.C.M. Plain Stem. ⅞ x 5" long. .... Price, each

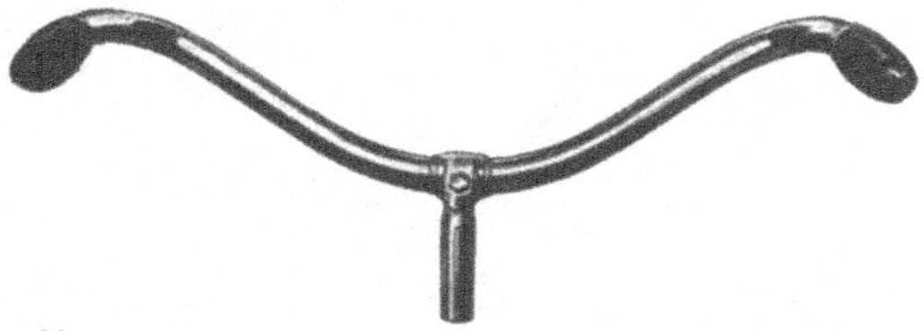

No. 953. *(Made in Our Own Factory)*

No. 953—C.C.M. Regular (24" Top) H. Bar and Post, less Grips .... Price, each

No. 960—24" H. Bar Top .... Price, each

No. 958—C.C.M. Plain Stem, ⅞ x 6½" long. .... Price, each

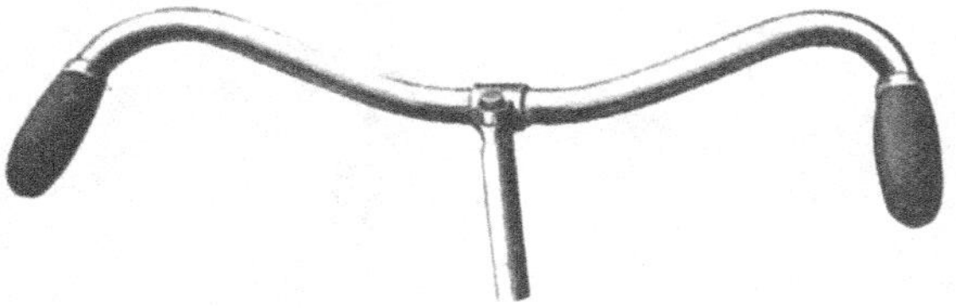

No. 965.

No. 965—Comfort H. Bar and Post, complete less Grips .... Price, each

No. 966—Comfort Top, 20" .... Price, each

No. 979—Comfort Plain Stem, ⅞ x 5¼" .... Price, each

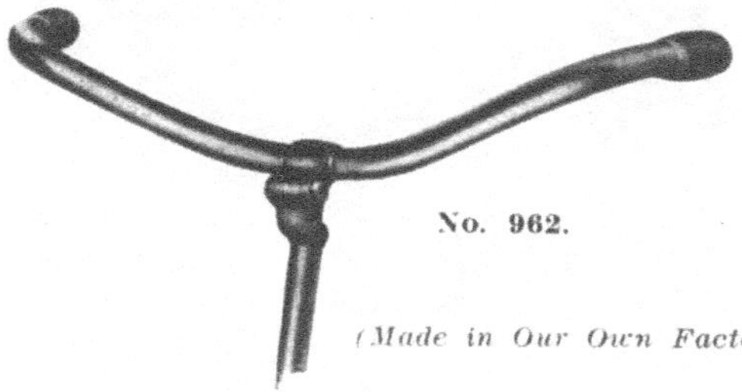

No. 962.

(Made in Our Own Factory)

No. 962—Sills Hygienic H. Bar and Post, less Grips .... Price, each

No. 956—20" Top .... Price, each

No. 961—Sills Hygienic Stem, ⅞ x 5½" long. .... Price, each

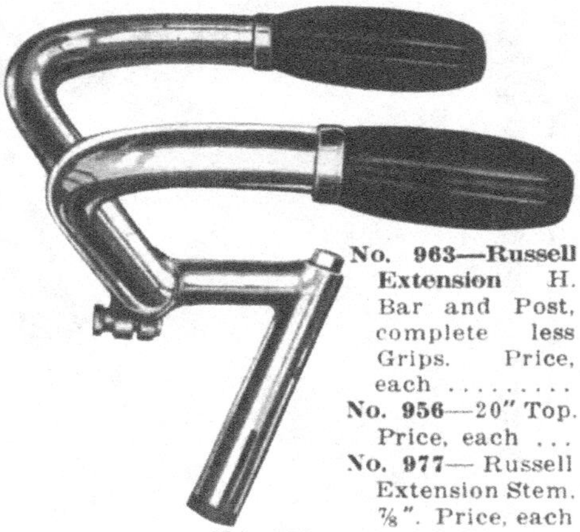

No. 963.

No. 963—Russell Extension H. Bar and Post, complete less Grips. Price, each ....

No. 956—20" Top. Price, each ...

No. 977—Russell Extension Stem. ⅞". Price, each

No. 968.

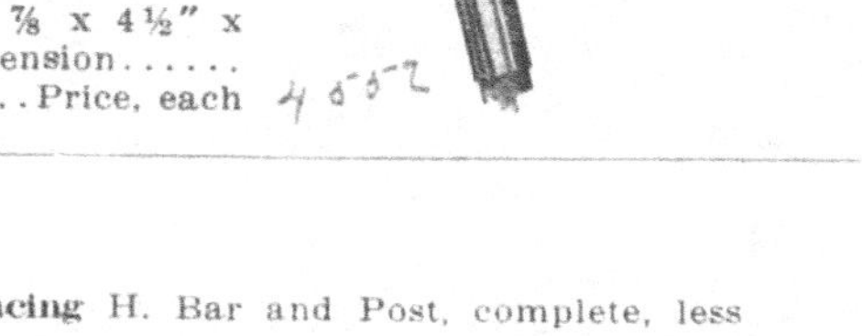

No. 968—Comfort Extension H. Bar and Post, complete, less Grips .... Price, each

No. 966—Comfort Top, 20" ...... Price, each

No. 970—Comfort Extension, ⅞ x 4½" x 2¾" extension...... .... Price, each

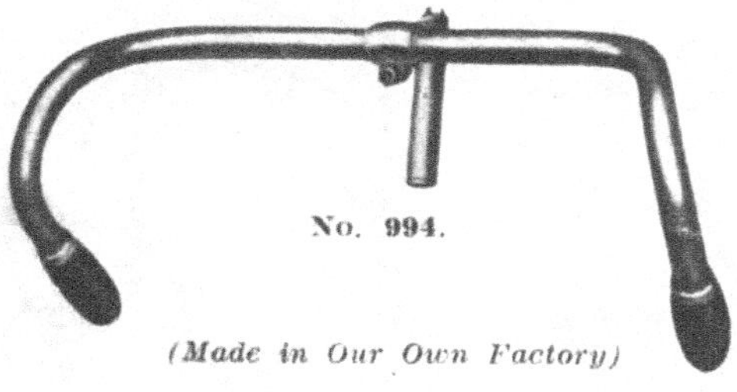

No. 994.

(Made in Our Own Factory)

No. 994—C.C.M. Racing H. Bar and Post, complete, less Grips .... Price, each

No. 974—(25)—17" H. Bar Top .... Price, each

No. 977—Russell Extension Post, ⅞" .... Price, each

"The Dealers' Interests Are Our Interests."

Handlebars (Continued)

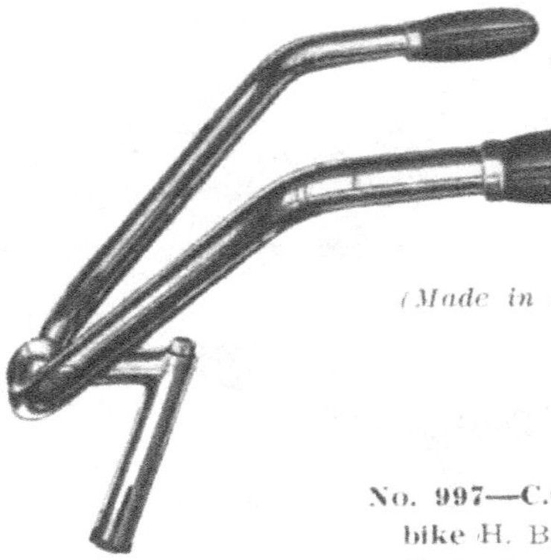

(Made in Our Own Factory)

No. 997.

No. 997—C.C.M. Motorbike H. Bar and Russell Extension Stem, complete, less Grips .... Price, each

No. 976—(27)—20" Top .... Price, each

No. 977—Russell Extension, ⅞" .... Price, each

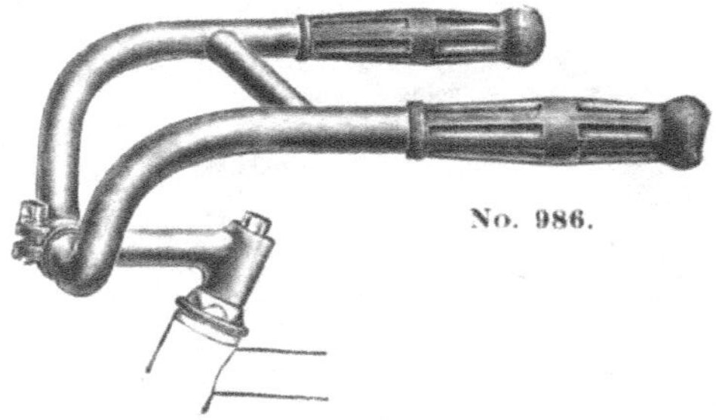

No. 986.

No. 986—"Motorbike Junior" H. Bar and Extension Stem, less Grips .... Price, each

No. 987—"Motorbike Junior," Top only. Price, each

No. 988—Stem only, ⅞". Length 4". Extension 4¼" forward, 1" downward .... Price, each

Handlebar Posts

(Made in Our Own Factory)

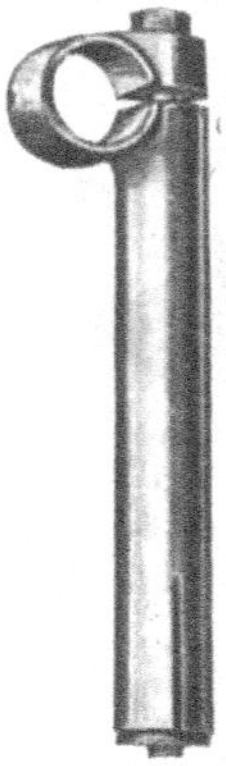

C.C.M.

No. 957—C.C.M. plain, ⅞ x 5". Price, each.

No. 958—C.C.M. plain, ⅞ x 6½". Price, each ......

Russell.

No. 977—Russell Extension, ⅞"... Price, each

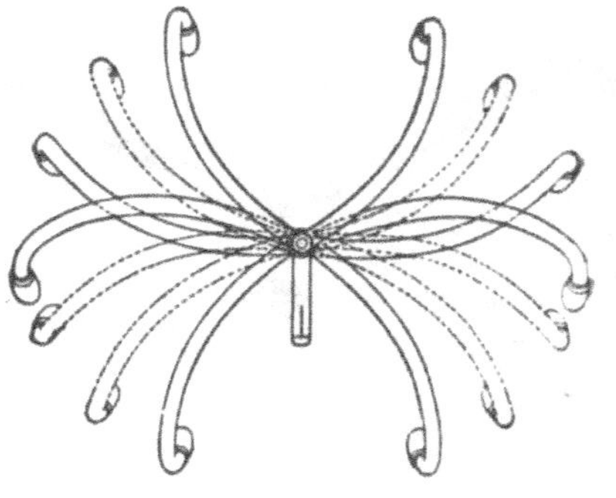

No. 969—Kelly Extension H. Bar and Post, complete, less Grips ....

B8—(Double row teeth) right side arm.. Price, each

B8a—(Single row teeth) left side arm.. Price, each

No. 996—Kelly 2½" Extension Stem... Price, each

Gooseneck.

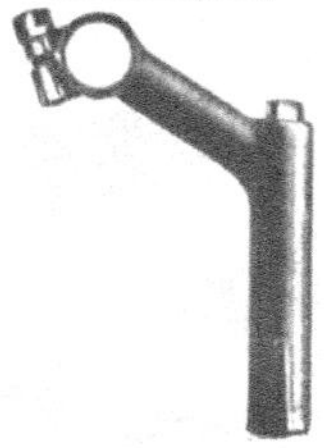

No. 985—Gooseneck Extension, stem ⅞". Price.. Each

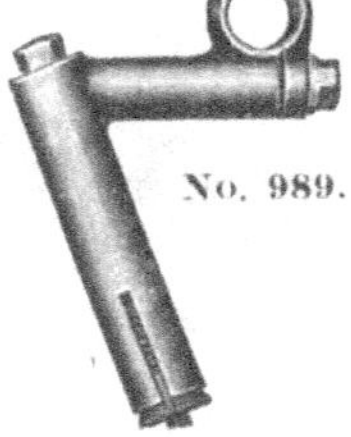

No. 989.

The new Extension Stem, which promises to be popular.

No. 989—Extension, ⅞" .... Each

Motorbike Jr.

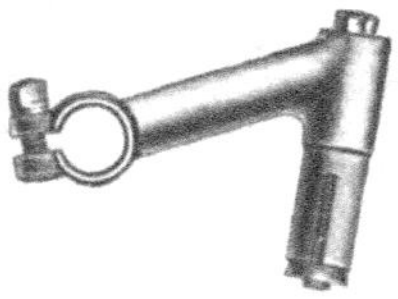

No. 988—"Motorbike Jr." Stem, ⅞ x 4 x 4¼" forward, 1" downward. Price, each ....

Comfort.

No. 970—Extension, ⅞ x 4½ x 2¾" extension. Price, each. **Comfort.** ....

No. 984—Extension, ⅞ x 9 x 2¾" extension. Price, each ....

COMFORT. PLAIN.

No. 979—Plain, ⅞ x 5¼".... Each

No. 967—Plain, ¾ x 5¼".... Each

No. 978—Plain, 13/16 x 5¼".... Each

No. 971—Plain, 1 x 5¼".... Each

No. 972—Plain, ¾ x 9". .... Each

No. 982—Plain, 13/16 x 9" .... Each

No. 983—Plain, ⅞ x 9". .... Each

KELLY STEMS.

No. 996—"Kelly" 2½" Extension Stem. Price, each ....

"Kelly Stem Parts.

B2—Forward Extension Stem Nut .... Each

B3—Forward Extension Stem Washer .... Each

B4—Forward Extension Stem Expander Screw. .... Each

B5—Forward Extension Stem Expander Plug. .... Each

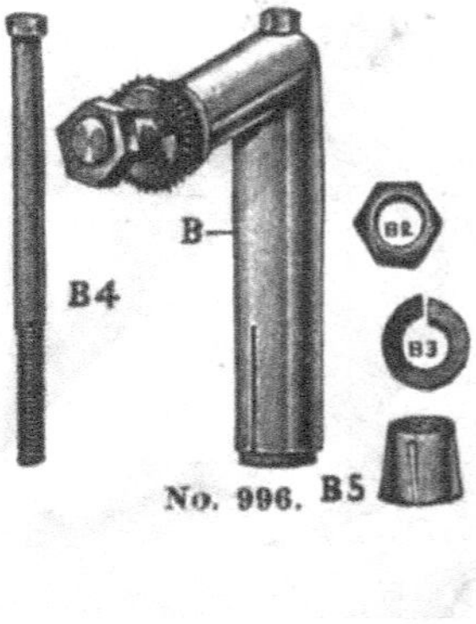

No. 996. B5

Prices Subject to Change Without Notice.

Handlebar Post

(Made in Our Own Factory)

No. 964.

No. 964—Sills' Hygienic, ⅞" x 5½" .... Each

EXTRA PARTS Sills' Hygienic Handlebar Post.

Post (stripped) .... Each
Hinge Clamp .... Each
Hinge Clamp Bolt .... Each
Hinge Clamp Bolt Nut.... Each
Hinge Bolt .... Each
Hinge Bolt Nut .... Each
Hollow Bolt .... Each
Expander .... Each
Plunger .... Each
Spring .... Each

Handlebar Bolts (with Expander Plug)

Bolts (Handlebar) with Expander Plug—These Bolts are manufactured from carefully selected steel, and are nicely finished. They have fine hexagon nickelled heads, and can be supplied in the following lengths:

Handlebar Tops

	Width.	Forward Bend.	Raise or Drop.	Price. Each
No. 956—(C.C.M.)	20"	2"	4"	
No. 960—(24)	24"	2"	5"	
No. 974—(25)	17"	1¾"	5¾"	
No. 976—(27)	20"	0"	6"	

NOTE.—All of the above tops are made in our own Factory.

No. 966—Comfort	20"	1¼"	3"	
No. 987—"Motorbike Jr."	20½"		4"	

No. 286—4"
No. 287—5" } Dozen........................
No. 288—6"
No. 289— 7"
No. 290— 8"
No. 291 – 9" } Dozen........................
No. 292—10"
No. 293—12" Dozen ....
No. 307—Expander Plugs, per dozen..........

Hangers

C.C.M. Crank Hanger

(Model "T," 1917 and 1918.)

The Crank Hanger is the heart of a bicycle, and it is therefore essential that it be designed on correct principles and carefully tooled and machined.

C.C.M. Cranks are drop forgings. Crank Hanger Cups and Cones are turned from bar steel, and ball races are ground perfectly true and polished. The machining is carefully and correctly done, resulting in perfect adjustment. The Cranks fit on the axle on a slight tapered bearing on three flat sides. This Hanger is regular equipment on C.C.M. Bicycles.

No. 1051—Price, each

C.C.M. Crank Hanger.

(Made in Our Own Factory)

Extra Parts For C.C.M. Crank Hanger

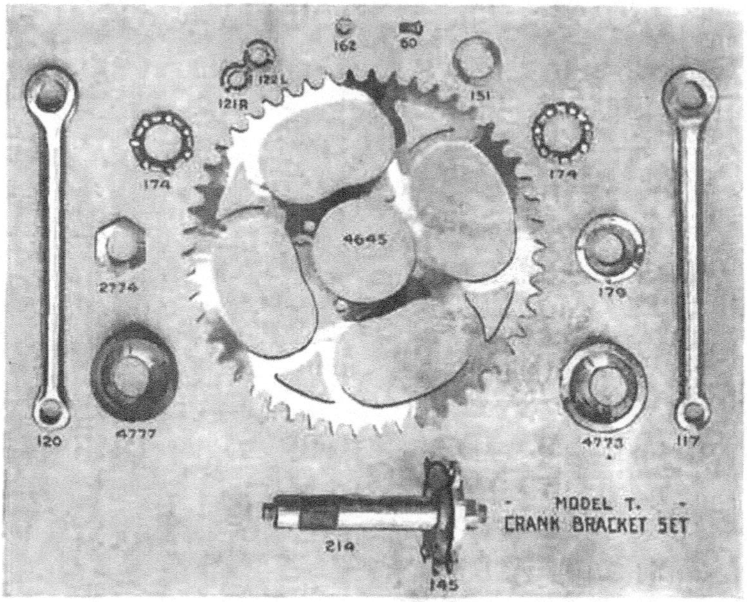

(Made in Our Own Factory)

1916, 1917 and 1918 Models.

No. 214—Axle used 1916/17/18 .... Each
No. 145—Stationary Cone used 1916/17/18 .... Each (Stationary Cone assembled on axle. See cut.)
No. 179—Adjusting Cone used 1916/17/18 .... Each
No. 174—Ball Cages (322) less balls ($\frac{3}{16}$ ball used), used 1916/17/18 .... ~~Doz.~~
No. 2774—Lock Nut, used 1916/17/18 .... Each
No. 151—Lock Nut Washer, used 1916/17/18 .... Each
No. 117R.—Right-hand Crank, used 1916/17/18 ....
No. 120L.—Left-hand Crank, used 1916/17/18 ....
No. 121R.—Right-hand Crank Jamb Nut, used 1916/17/18 ..... Each
No. 122L.—Left-hand Crank Jamb Nut, used 1916/17/18 ..... Each
No. 4645—Sprocket, 50 tooth, ½" pitch, ⅛" wide, used 1916/17/18 .... Each
No. 60—Sprocket Bolt, used 1916/17/18 .... Doz.
No. 162—Sprocket Bolt Nut, used 1916/17/18 .... Doz.
Steel Balls—5/16, used 1916/17/18. .... Per gross
Coupler and Ball Cases (ball bearing sleeve), used 1916 only (not shown in plate) .... Each
Ball Case Collar—(Ball Retainer washer), used 1916 only (not shown in plate) .... Each
No. 4777—Right Cup (black finish) used 1917 and 1918 only, model T .... Each
No. 4773—Left Cup (nickel plated finish), used 1917 and 1918 only, model T. .... Each

HANGERS, Extra Parts for Hercules Crank Hanger

For replacement on Model "F" Bicycles.

Cranks and Axle (one-piece) ....Each
Stationary Cone ....Each
Adjusting Cone ....Each
Lock Nut ....Each
Lock Nut Washer ....Each
Ball Cups ....Each
Ball Cage (No. 321), less balls ....Doz.

(Made in Our Own Factory)

Balls, 5/16" ....Gross
Dust Shield ....Each
Sprocket, 50 x ½ x ⅛" ....Each
No. 1911—Special Stationary Cone Spanner for "Hercules" Hanger ....Each
Width of Hanger ....2 31/64 inches
Outside diameter of Ball Cups ....1 61/64 inches

HANGER, Stevens' Repair

The only Hanger which will fit absolutely every size bracket. One size does it all.

This is a two-piece Hanger, with shaft milled flat on one side long enough to reach through any old style hanger.

The crank shaft and left hand crank are threaded. After adjusting the cones, **screw left crank** up against lock nut, put in cotter pin and **saw off the shaft** to desired length. All sizes of cups and cones to fit same are carried in stock.

In ordering, it is advisable to send sample of bracket cup, or, where that is not possible, the exact inside diameter of the crank bracket.

Cones—**No. 551/2** supplied with Group No. 1, **No. 533/4** with Group No. 2, **No. 555/6** with Group No. 3, **No. 557/8** with Group No. 4, and **No. 559/60** with Group No. 5.

Price, complete with Cups (one in box).Each
Price, less Cups ....Each
Cups, Nos. **501-521** ....Per pair
Cones, Nos. **551-560** ....Per pair
11/16" x 24 **Lock Nuts** ....Each
11/16" **"D" Washers** ....Per 100
⅜" **Cotter Pins** ....Dozen
Sprockets, 3/16 x 24 and 26 tooth.....Each
Sprockets, ¼ x 24 and 26" tooth.....Each
Right Crank and Axle ....Each
Left Crank ....Each

Below we give a list of all the cups and cones and their sizes, and also state which cups each pair of cones will fit. Every cup and cone has a number. **Order by number.**

4709
4703
4547
4546
4548
4548
4546
4541

CUPS.

	Cup Number.	Outside Diameter.	Size of Balls.
Group No. 1	501	1 1/2 in.	1/4
	502	1 17/32 in.	1/4
	503	1 9/16 in.	1/4
Group No. 2	504	1 19/32 in.	5/16
	505	1 5/8 in.	5/16
	506	1 21/32 in.	5/16
	507	1 11/16 in.	5/16
	508	1 23/32 in.	3/8
Group No. 3	509	1 3/4 in.	3/8
	510	1 25/32 in.	3/8
	511	1 13/16 in.	3/8
	512	1 27/32 in.	3/8
	513	1 7/8 in.	3/8
Group No. 4	514	1 29/32 in.	3/8
	515	1 15/16 in.	3/8
	516	1 31/32 in.	3/8
	517	2 in.	3/8
Group No. 5	518	2 1/32 in.	3/8
	519	2 1/16 in.	3/8
	520	2 3/32 in.	3/8
	521	2 1/8 in.	3/8

CONES.

Cone Number	Outside Diameter.	Inside Size.	Fits Group Number
551	1 3/8 in.	11/16x24 L.H.	1
552	1 3/8 in.	3/4x24 R.H.	1
553	1 7/16 in.	11/16x24 L.H.	2
554	1 7/16 in.	3/4x24 R.H.	2
555	1 5/8 in.	11/16x24 L.H.	3
556	1 5/8 in.	3/4x24 R.H.	3
557	1 3/4 in.	11/16x24 L.H.	4
558	1 3/4 in.	3/4x24 R.H.	4
559	1 13/16 in.	11/16x24 L.H.	5
560	1 13/16 in.	3/4x24 R.H.	5

All the cups in any one group are of the same **inside** diameter, and therefore all cups of one group are fitted by one pair of cones. It takes only the five pairs of cones to fit the 21 sizes of cups. There are 21 outside diameters to the cups, but only five **inside** diameters.

Head Fittings

(For all C.C.M. High Grade Bicycles.)

4708—The following parts made up complete ....Per set
4709—Adjusting Cone Lock Nut....Each
4703—Adjusting Cone Lock Washer.Each
4547—Upper Adjusting Cone .....Each
4546—Ball Cage (5/32 balls used) upper or lower (less balls)......Each
4548—Upper or lower cup ....Each
4541—Stationary Cone ....Each

Holders, Bicycle Wall

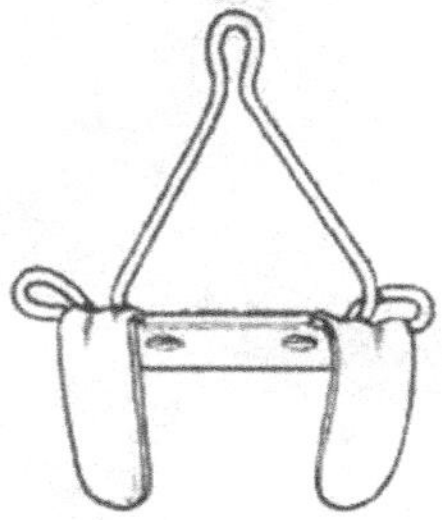

Bedley Top.
(Made in Canada)

Bedley—This holder is made from galvanized wire, the top holder being leather covered. It is furnished complete with brace and screws.
No. 990—Per dozen . .

Bedley Bottom.
(Made in Canada)
"Bedley Bottom" is a plain wire holder not covered.
No. 992—Per dozen . . .

The "Gem" is a plain holder leather covered. Screws supplied.
No. 991—Per dozen . .

Gem.
(Made in Canada)

Horns, Bicycle

(We can supply parts for Horns listed.)

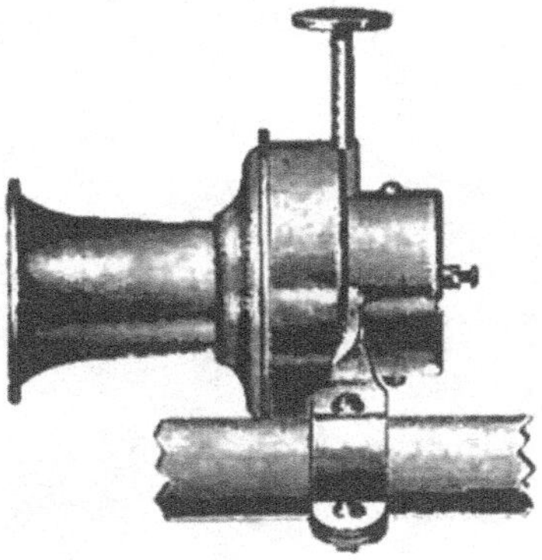

Bevins Long Distance

Hand operated Horn, substantially made, loud, penetrating tone. Black enameled, with nickel-plated bracket.

No. 1048—Without Projector, all Black. . . . . Each

No. 1049—With Projector, all Black. Each

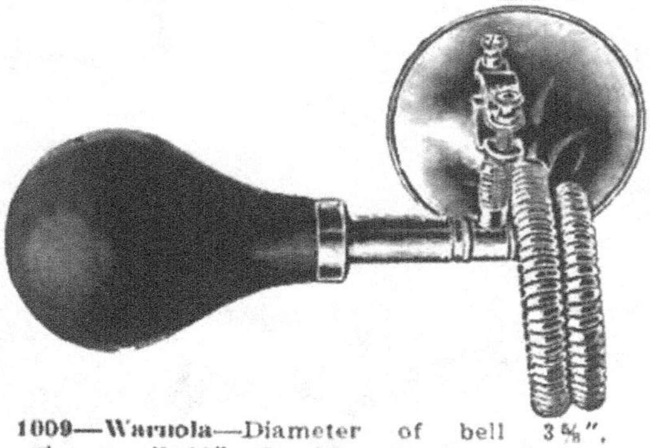

No. 1009—Warnola—Diameter of bell 3⅝", length over all, 10". Bends made of stiff tubing, not flexible as illustrated. Fine deep tone. Nickel-plated. No. 5 Bulb used . . . . Each
No. 1009R—Reeds for above . . . . Each
No. 1009B—Bulbs for above . . . . Each

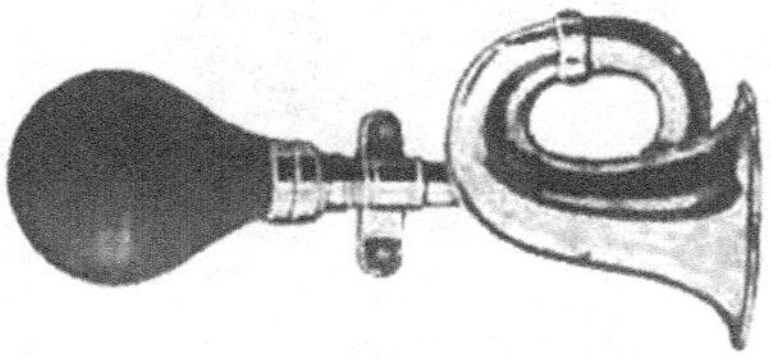

No. 999.

No. 999—(26) Bicycle Horn, diameter of bell 2¾", length over all 8¾", fixed clip, highly nickel-plated . . . . Each

No. 1015—Reeds for above . . . . Each

No. 1016—Bulbs for above . . . . Each

Starr

No. 1052.
Medium priced effective Warning Signal, black enamel finish, complete with attaching bracket.
No. 1052—(One in a carton). Each

"Screech Owl"

No. 1093—"Screech Owl." A very superior "Siren," nickel-plated. Fitted with ball bearings. (2 in box) . . . . . . Each

Whangdoodle

No. 998—Whangdoodle Bicycle Vibrator Horn.
Black enamel, with nickel-plated chain. (One in box). Price, each . . . .

Motorcycle Horns

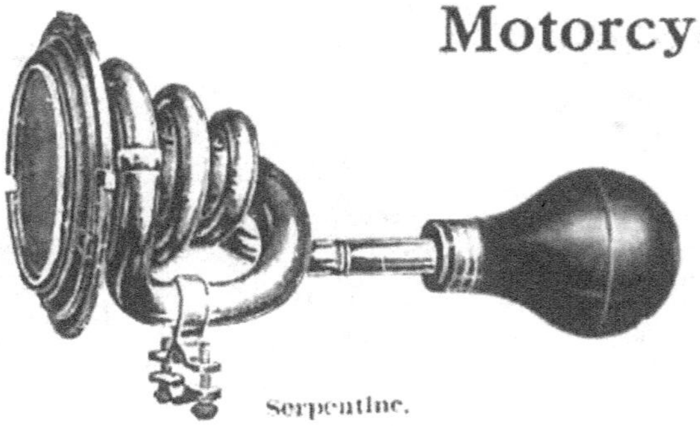

Serpentine.

Specially recommended for motorcycle use. Very deep note. Finest nickel finish. Diameter of bell, 3⅞", length over all, 12". (One in carton.)

No. 1007 (92) .... Each
No. 1007R—Reeds .... Each
No. 1007B (6)—Bulbs .... Each

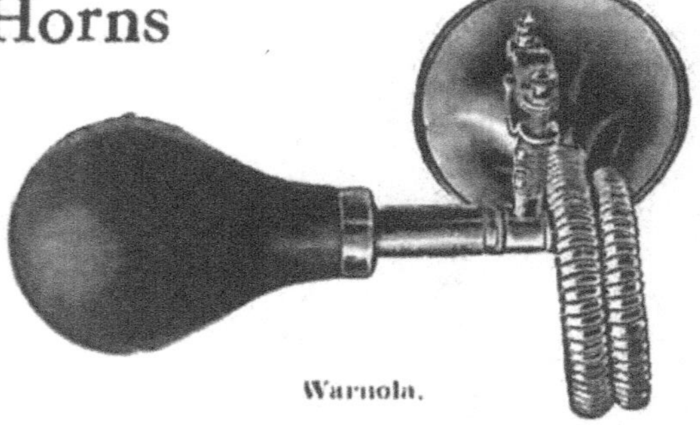

Warnola.

Diameter of bell, 3⅝". Length over all, 10". Fine, deep tone. Bends made of stiff tubing, not flexible as illustrated. Nickel-plated. No. 5 bulb used.

No. 1009—(18B) .... Each
No. 1009R—Reeds .... Each
No. 1009B (5)—Bulbs .... Each

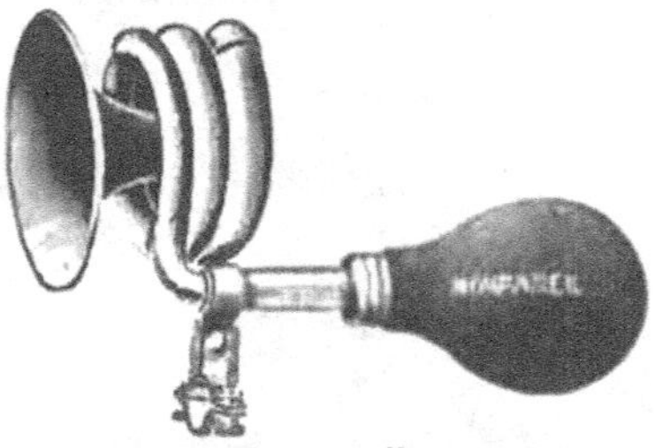

Nonparcell.

Fine nickel finish. Diameter of bell 4¼". Length over all, 10". (3 in box.)

No. 1020 (102)— ... Each
No. 1007R— Reeds ... Each
No. 1007B (6)— Bulbs ... Each

E. A.

Size 4 x 5½"; weight two pounds. Supplied with two brackets for attaching to handlebar or top bar of frame of Motorcycle. To meet the demand for a horn for Motorcycles finished in service green or khaki, this model can now be supplied in **Khaki as well as Black and Nickel.**

No. 1022—Black and nickel .... Each
No. 1023—Khaki Each

Hubs

Comfort Hubs

(Made in Our Own Factory)

Comfort—Front.

These are high grade Hubs, turned from best bar steel stock. Bearings are accurately ground after being hardened. Fitted with ball retainers and dust washers. Absolutely dust proof. Workmanship and finish are of highest quality. Fitted with sprockets. **36 spoke hole front hubs will be supplied unless otherwise ordered.**

These hubs are fitted with spring ball type oiler.

No. 1036—(Comfort) Front, 36 hole .... Each
No. 1032—(Comfort) Front, 32 hole .... Each
No. 1030—Rear .... Each

(Made in Our Own Factory)

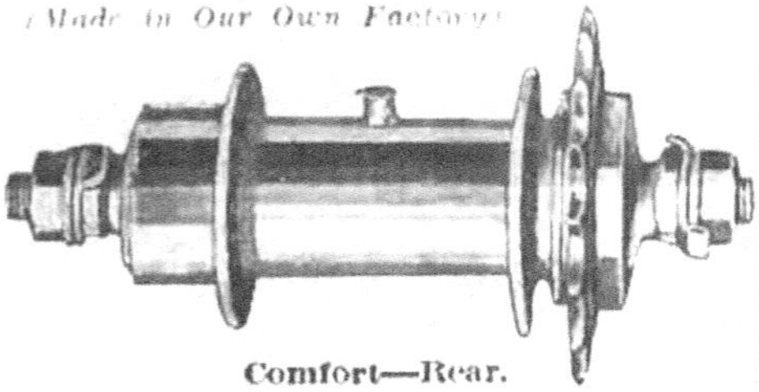

Comfort—Rear.

(Made in Our Own Factory)

No. 25 Front Hubs—High grade pressed Hub. Barrel made of tubing, with ends pressed on. Nicely plated. 32 and 36 hole

No. 1047—Drilled 36 Each

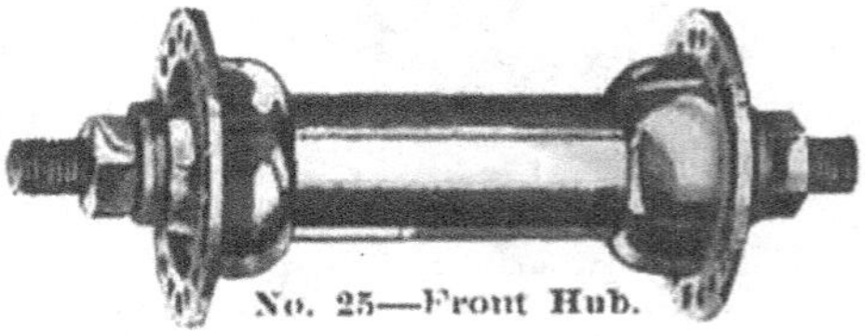

No. 25—Front Hub.

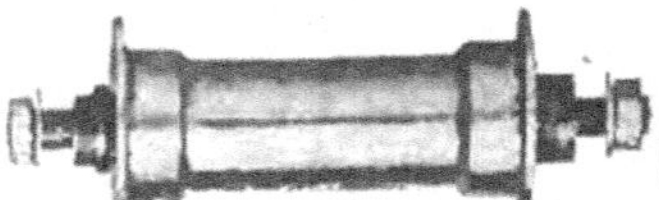

Sulky Hubs

(Made in Canada)

The Knockout Axle Hub is so much more satisfactory that we have discontinued the plain axle hub.

No. 1040—A superior hub with special features and finish. Has knockout axle, which makes it easy to keep clean. Drilled for 36 spokes.

No. 1040 .... Each
No. 1041—Axle . Each
No. 1042—Cone . Each
No. 1043—Cup .. Each
No. 1044—Axle Sleeves .... Each

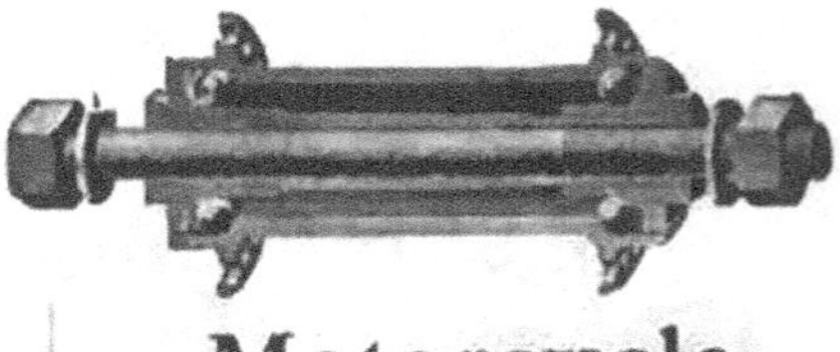

Motorcycle Front Hubs

No. 1045—Drilled 36 .. Each
No. 1046—Drilled 40 .. Each

Hub Parts

See page 13.

Lamps

It Pays to Push Lamps

A GOOD LAMP for the Bicycle is a **necessity** rather than a **luxury.** Many Towns are passing By-laws compelling Bicycles as well as other vehicles to carry a light. Even where there is no legal compulsion, public opinion is strongly in favor of the use of lamps as a "Safety First" measure.

This means that it will pay the Bicycle dealer well to push strongly for lamp business.

In the "Old Sol" line we offer Dealers a wide range of well made, attractive lamps for both Bicycles and Motorcycles. In spite of the extremely high cost of both labor and materials **the prices have been considerably reduced** and "Old Sol" lamps are without doubt the best value to be had at the present time.

Bicycle Oil Lamps

Kingfisher — A well-made, reliable English lamp of good size and appearance. Brass nickel-plated; fitted with a wind-proof head and convex lens; complete with bracket and red side jewels. Side jewel hinged. **Burns C.C.M. Burning Oil, not kerosene.**

No. 1122 (25)—(One in box) ..... Each

Kingfisher.

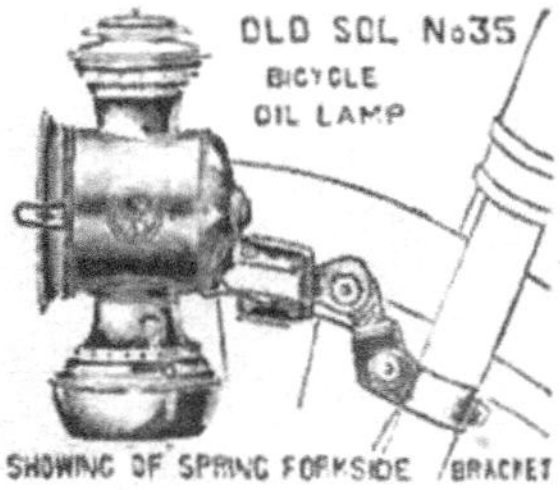

Equipped with Spring Shock Absorbing Bracket. Made from brass, nickel finish. Burns about 14 hours on one filling. Red and green side jewels. Polished reflector. Height 6¾", diameter of door 3⅜", weight 1 lb. 2 oz. Complete with bracket. **Burns kerosene (coal oil).**

No. 1119 (35)—(One in box) ....Each

Bicycle Carbide Lamp

Panther—A good English lamp at a low price. Made of all brass, nickel-plated and highly polished. Burns with a steady light. Complete with bracket. (One in box.)

No. 1125 (20)—(No 1153 Burner used) ...... Each

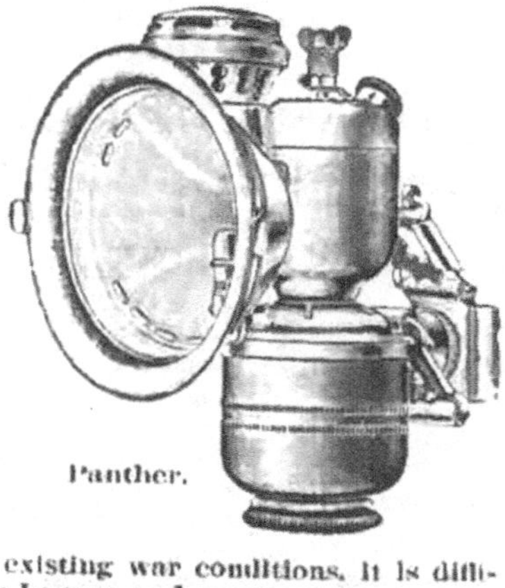

Panther.

Notice—Owing to existing war conditions, it is difficult to secure Panther Lamps and we cannot guarantee delivery.

"Old Sol" Bicycle Carbide Lamps

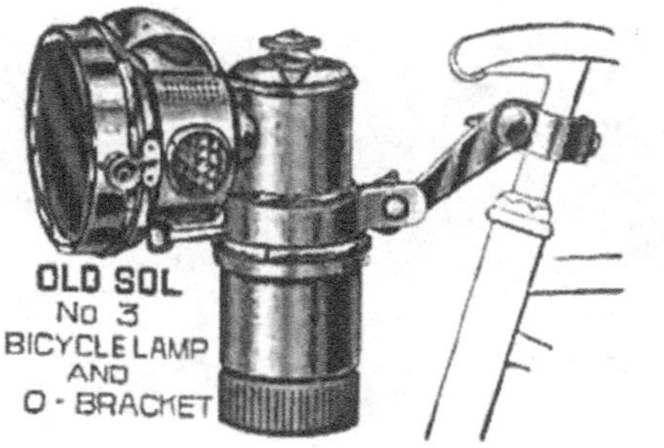

Large, handsome Bicycle Lamp of motorcycle type. Cannot blow out.

Height 7". Carbide capacity 6 oz. Diameter of door 4". Burns 8 hours on one filling. Heavy brass, nickel-plated, with black enamelled Bracket. Red and green side jewels. Polished reflector.

No. 1141 (3)—Price with Bracket (one in box) .... Each

No. 1140 (10)

Safe, attractive gas lamp, which requires very little care. The improved process generates gas at regular intervals. Will not leak or corrode.

Height 6¼". Carbide capacity 5 oz. Diameter of door 3¼". Burns 6 to 7 hours on one filling. Heavy brass, nickel-plated, with black enamelled bracket. Red and green side jewels. Polished reflector.

No. 1140 (10)—Price with bracket (one in box) ....Each

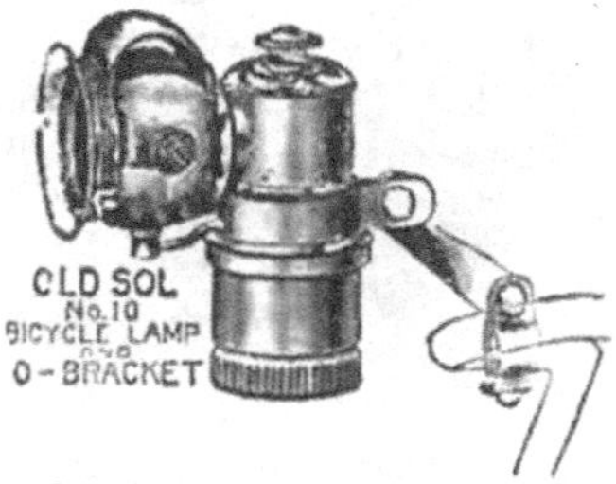

Showing Red Rear Signal.

No. 1142 (23)

A powerful front white light and a red rear signal, which is a protection to rider from vehicles approaching from rear. For use on carriages and other vehicles.

Height 6½". Diameter of door 3⅜". Carbide capacity 5 oz. Burns 6 to 7 hours. Heavy brass, nickel-plated, with black enamelled Bracket. Red side jewel. Polished reflector.

No. 1142 (23)—with Bracket (one in box) ....Each

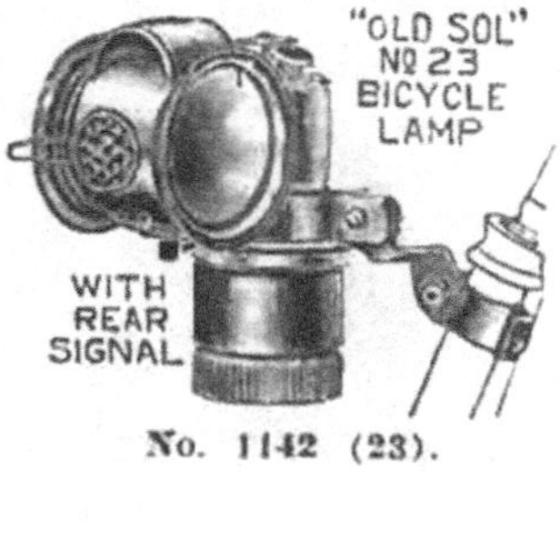

No. 1142 (23).

Lamps (Continued)

"Old Sol" Gas Lamps, Motorcycle

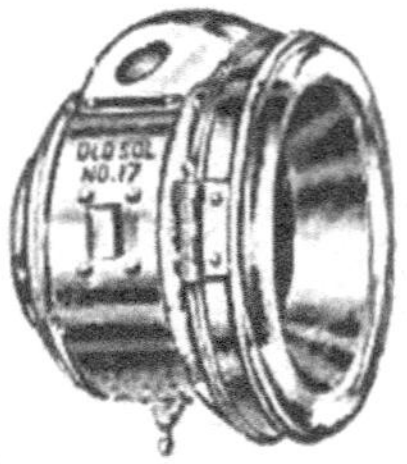

Bodies made from one-piece heavy gauge brass, nickel-plated. Door fitted with ball pin hinges. Front diameter 7½", height 7¾", width 8¼", depth 5½". Genuine 5" Mangin mirror lens. Distance between props 6½". Fits all Old Sol Standard Brackets. Regular ½ ft. burner used. **Without bezel shown in illustration.**

No. 1164 (17)—Black and Nickel, with A. or C. Brackets. (1 in box). Price, each

No. 1165 (17)—Nickel, with A. or C. Brackets. (1 in box) ....Price, each

Identical in size with No. 1164 (17). Fitted with rear socket for spade bracket. Red and green side jewels. Regular ½ ft. burner used. **Without bezel shown in illustration.** Takes 5" lens.

No. 1161 (16)—Black and nickel (less bracket) (one in box) ....Each

No. 1162 (16) — Nickel (less bracket) (one in box) ....Each

No. 1161 (16).

"Old Sol" Gas Lamps, Motorcycle

No. 1166 (8).

The leader of the "Old Sol" line. Front diameter 8½", height 9", width 9¼", depth 6½". Fitted with genuine 6" Mangin mirror lens. Distance between props 7½". Fits all "Old Sol" standard brackets. Regular ½ ft. burner used. **Without bezel shown in illustration.**

No. 1166 (8)—Black and nickel, with A. or C. bracket (one in box) .....Each

No. 1167 (8)—Nickel, with A. or C. bracket (one in box)..Each

Popular for all classes of motorcycles. Front diameter 6½". Genuine 4½" Mangin mirror lens. Height and width 7½". Depth 5¼". Equipped with rear socket for spade bracket. **Without bezel, as shown.** ⅜ ft. burner used.

No. 1163 (2)—Nickel only (less bracket) (one in box) ....Price, each

A splendid **side-car lamp**, having red rear signal and red and green side jewels. Door 4". Polished reflector. 1/6 ft. burner used. **No bezel.**

No. 1168 (13)—Nickel, complete with Bracket. (One in box)... ....Price, each

"Old Sol" Electric Lamps, Bicycle

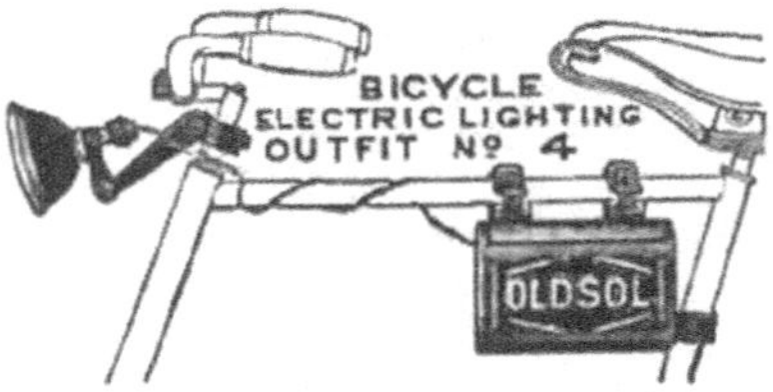

Will appeal to many boys who have been looking for a real electric lamp at a reasonable price. Has large attractive headlight with bracket for handlebar.

Lamp all black enamelled. Uses two regular No. 6 batteries. Supplied with bulb 2.8 volts 2 C.P. double contact, Ediswan base. Face of lamp 5" diameter. Lamp, Wiring and Bracket packed in one box. Dry Cell Container in another box.

No. 1144 (4)—(Less Battery) ......Price, each

High grade dependable electric lamp, which will satisfy any rider wishing a Dry Cell outfit.

This lamp has black enamelled body, with nickel-plated door. Uses two regular No. 6 batteries. Supplied with bulb 2.8 volt, 2 candle-power, single contact. Ediswan base.

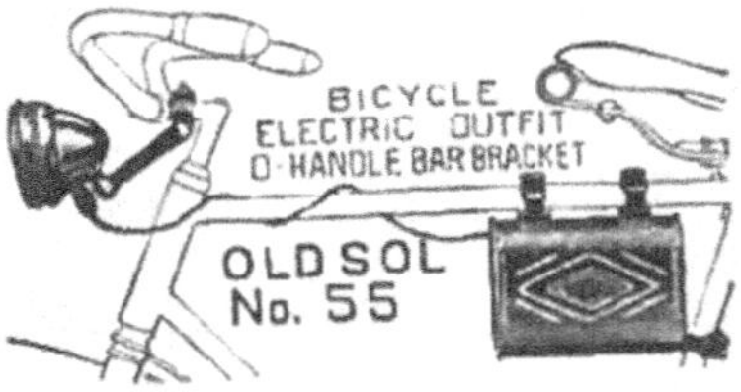

Diameter of door 4½". Highly polished silver reflector.

Lamp, Wiring and Bracket packed in one box. Dry Cell Container in another box.

No. 1145 (55)—(Less Battery) ......Price, each

"Old Sol" Electric Side Car Lamp

Without a headlight on the side car you are in danger of being struck by an automobile or vehicle, as it is difficult for the driver to judge the space you require. With a lamp on the side car and one on the motorcycle you will secure as much room as an automobile.

Very neat bullet shaped lamp, 4" in diameter.

No. 1173 (40)—Black and nickel, 6 V., 2 C.P., double contact, Ediswan bulb used..... ....Price, each

"Old Sol" Tail Lamp, Motorcycle

Acetylene Gas

"OLD SOL" NO. 15 TAIL LAMP

Substantial Gas Tail Lamp and License Holder, for rear guard of motorcycle. Red glass 2⅝" diameter; ⅛ foot burner used. Wire screen in base of lamp for illuminating License Bracket and for ventilation.

Off and on valve in gas tube for regulation of gas. Bracket with side arms for License Bracket, and holes with bolts and nuts for attaching to mudguard. Red or grey finish.

No. 1169 (15)—Service Green, for Harley-Davidson and Excelsior Motorcycle (one in box) ....Price

No. 1170 (15)—Red (one in box) ....Price

"OLD SOL" No. 61 TAIL LAMP

"Old Sol" Electric Tail Lamp

Has 3" rear red signal and license plate bracket, with bolts and nuts. White celluloid in base of lamp to illuminate license plate. 6 V., 2 C.P., single contact, Ediswan base bulb used. Black finish.

No. 1171 (64)—(One in box) ....Price, each

OLD SOL No. 5 GENERATOR WITH E BRACKET

Carbide Generator

Capacity 12 oz. carbide, one pint of water, will burn "Old Sol" No. 8 for about 6 hours. Height 9", diameter 4⅞". All nickel finish. Carbide chamber 4" high, 3" diameter.

No. 1186 (5)—With E. or H. Bracket ....Price, each

No. 1198—E Bracket for Frame ....Price, each

No. 1199—H. Bracket for Front Fork ....Price, each

No. 1150.

Lamp Brackets

Made of heavy sheet metal and spring steel wire, heavily nickeled and polished, light and strong, adjustable, felt lined.

No. 1150 — (12 in box) ..Per dozen

Lamp Brackets, Carriage

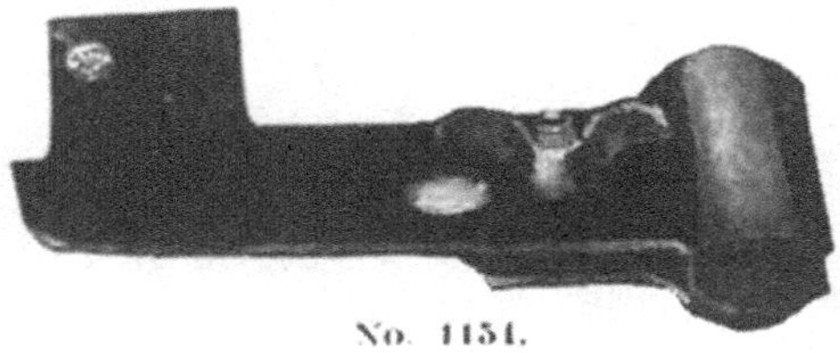

No. 1151.

(Made in Our Own Factory)

Made from heavy sheet steel, japanned. This bracket is fitted with an adjustable clamp, easily attached to dash board.

No. 1151—Price ....Each.

"Old Sol" Motorcycle Lamp Brackets

U. B. T. No. 5 Fork Side Clamps

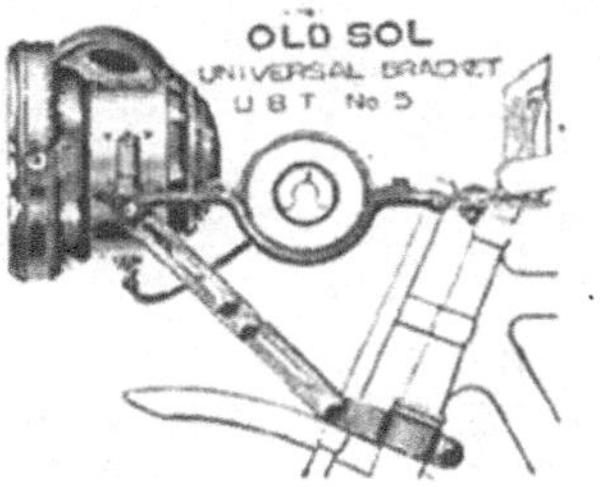

Carries Prest-O-Lite Tank and Lamp. Adapted for Harley-Davidson, Henderson, Reading-Standard, Excelsior. Enamel and nickel finish.

No. 1197 (5)—(One pair in box).Price

Lamp Burners, Acetylene Gas

No. 1152.

No. 1155—Solar Burners for Solar Gas Lamps ....Price

No. 1152—Lava Tips. ¼ foot. Can be used for either Solar or 20th Century Gas Lamps. 1 gross in box ....Per dozen

No. 1155

Bracket "A"

Can be fitted in front of fork head or behind it, as shown in illustration. Will fit practically all motorcycles, including Indian, Henderson, Reading-Standard, Sears, Excelsior, Iver Johnson, Thor, Flanders, Pope, Yale, Pierce, Harley-Davidson, etc.

No. 1194 (A)—Price ..Per pair

Bracket "A" *Behind Head*

"Old Sol" Burners

No. 1206 (OS106)—1/6 foot, for "Old Sol" Bicycle Carbide (Gas) Lamps .....Each

No. 1207 (OS200)—½ foot, for "Old Sol" Motorcycle Carbide (Gas) Lamps Nos. 1161 (16), 1164 (17), and 1166 (8).Each

No. 1208 (OS39)—⅜ foot, for "Old Sol" Motorcycle Carbide (Gas) Lamps Nos. 1163 (2) and 1168 (13) ....Each

Bracket C

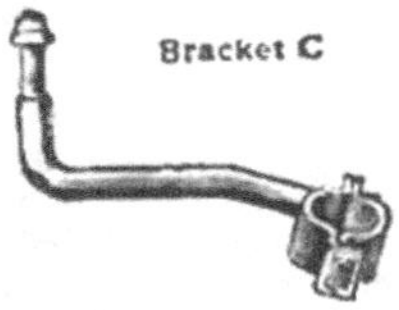

Bracket "C"

Will fit Harley-Davidson, Pope, Yale, Emblem, Merkel, Excelsior, etc.

No. 1195 (C)—Price, per pair

P. & H.

This burner gives a flat flame from a single gasway. It can be easily cleaned. It is used on all Powell and Hanmer Bicycle gas lamps.

No. 1153—Price, doz. (1 gross in box)

No. 1153.

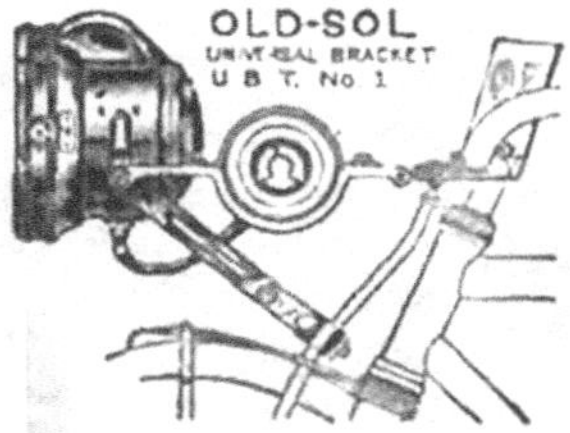

U. B. T. No. 1 Truss Clamps

Carries Prest-O-Lite Tank and Lamp. Adapted for Indian, Excelsior, Pope, Thor, Merkel, Dayton. Enamel finish.

No. 1196 (1)—Price

Bray's Deto

No. 1154—Price, doz. (1 gross in box) ...

Lamp Parts

Searchlight Lamp Parts

No.	Description.	Price Each.
1	Top	$2.60
2	Top Cap	.40
3	Reflector, complete	.40
4	Screw and Washer for Reflector	.14
5	Window and Wires	.48
6	Lens	.42
7	Lens Wire	.14
8	Fount	2.80
9	Filler Cap	.30
10	Hanging Tube Cap	.20
11A	Burner with Metal Collar and Lava Tip	.32
12	Presser Spring	.20
13	Carbide Cup	1.00
14	Bracket	1.00
15	Head Clamp	.40
16	Bracket Clamp	.40
17	Bracket Set Screw	.14
18	Bracket Adjusting Screw	.14
19	Head Clamp Screw and Wing Nut for Bracket	.30
20	Rubber Gasket	.20
22	Carbiot Gas Strainers	.08

NOTE.—Always specify for Searchlight Gas Lamp when ordering above parts.

Powell & Hanmer Lamp Parts

Panther		No. 1125
Panther Self-Lighter. (Not Listed)		No. 1126
Panther Duplex. (Not Listed)		No. 1128
Revenge (Not Listed)		No. 1129
P. & H. (Not Listed)		No. 1130
Demon		No. 1120
Referee. (Not Listed)		No. 1121
Kingfisher. (Not Listed)		No. 1122
P. 1—Glasses for Nos. 1125, 1126, 1128	Each	$0.24
P. 1A—Glasses for No. 1129	Each	.24
P. 2—Glasses for No. 1130	Each	.54
P. 3—Glasses for No. 1120 (with rims)	Each	.36
P. 4—Glasses for No. 1121 (with rims)	Each	.48
P. 5—Glasses for No. 1122	Each	.48
P. 6—Burner No. 1153. (Screw-in)	Doz.	1.60
P. 7—Water Caps	Each	.20
P. 8—Valves	Each	.36
P. 9—Valve Screws	Each	.24
P. 10—Zinc Tubes	Each	.24
P. 11—Filter Pads	Each	.24
P. 12—Large Rubber Washers	Each	.24
P. 13—Small Rubber Washers	Each	.10
P. 14—Adjustment Screws	Each	.24
P. 15—Springs and Plates	Each	.24
P. 16—Pad Plates and Nuts	Each	.40
P. 17—Back Springs	Each	.10
P. 18—Bottom Screws	Each	.40
P. 19—Carbide Chambers for No. 1125	Each	.40
P. 20—Carbide Chambers for No. 1130	Each	.52
P. 20A—Carbide Chambers for No. 1129	Each	.52
P. 21—Cross Gate	Each	.24
P. 22—Smoke Cap	Each	.24
P. 23—Cones for No. 1125	Each	1.40
P. 24—Cones for No. 1129	Each	2.10
P. 25—Cones for No. 1130	Each	2.80
P. 26—Pad Plates	Each	.24
P. 27—Backs	Each	.40
P. 28—Windups for Nos. 1120, 1121, 1122	Each	.24
P. 29—Founts comp. for Nos. 1120, 1121, 1122	Each	.40
P. 30—Blue Springs for No. 1120	Each	.10
P. 31—Wire for Glasses	Each	.10

NOTE.—When ordering any of above Lamp Parts, always give the number of lamp for which part is required.

"Old Sol" Lamp Parts

Below we list the most important replacement parts for "Old Sol" Bicycle and Motorcycle Lamps. Should you require other parts for these Lamps which are not listed, kindly send for special parts list and quotation on desired parts.

Always give number of Lamp when ordering parts.

Parts for No. 1119 (35) "Old Sol" Bicycle Oil Lamp.

OS463	Front Door Glass	$0.20
OS1216	Reflector with hole in centre	.50
OS1224	Burner, less wick	.20
OS1226	Wick only	.20
OS1531	Retaining Band for Door Glass	.20

"Old Sol" Lamp Parts—(Continued)

Parts for "Old Sol" 1164 (17) and 1161 (16) Motorcycle Gas Lamp.

No.	Description.	Price Each.
OS200	Burner ½ ft. Standard Pillar	$0.60
OS432	Door Glass	.30
OS433	Glass Retaining Ring	.30
OS436	Retaining Ring for Mirror Lens	.30
OS439	Burner Holder	.30
OS525	Mangin Mirror Lens, 5"	3.00
OS798	Bezel or Door Reflector	1.50

Parts for "Old Sol" 1166 (8) Motorcycle Gas Lamp.

OS30	Burner Holder	$0.30
OS193	Door Glass, 8"	.40
OS200	Burner ½ ft. Standard Pillar	.60
OS201	Mangin Mirror Lens, 6"	3.50
OS206	Retaining Wire for Door Glass	.30
OS641	Wire Retaining Ring for Mirror Lens	.30
OS799	Bezel or Door Reflector	2.00

Parts for "Old Sol" 1163 (2) Motorcycle Gas Lamp.

OS7	Retaining Ring for Door Glass	$0.30
OS9	Door Glass	.30
OS23	Burner Holder for Pillar	.30
OS39	Burner ⅜ ft. Standard Pillar	.60
OS41	Reflector 4½". Includes frame	3.00
OS797	Bezel or Door Reflector	1.50

Parts for "Old Sol" 1168 (13) Motorcycle Gas Lamp.

OS103	Door Glass	$0.30
OS106	Burner 1/6 ft., Standard Pillar	.40
OS530	Reflector with Hole	.50
OS1390	Retainer for Door Glass	.30

Parts for 1141 (3) "Old Sol" Bicycle Gas Lamp.

OS73	Reflector, New Style	$0.40
OS74	Carbide Retaining Disc only	.20
OS76	Carbide Chamber only	2.00
OS84	Water Filler Cap, with Leather Washer	.20
OS107	Carbide Basket	.60
OS90	Valve Rod, complete	.80
OS94	Spring for Carbide Plate	.10
OS98	Rubber Gasket	.20
OS100	Binding Screw for Head and Generator	.20
OS103	Door Glass	.30
OS106	Burner 1/6 ft. Standard Pillar	.40
OS398	Valve Seat, Knurled	.20
OS487	Spring Retainer for Filter Plate	.10
OS489	Felt Filter	.10
OS502	Filter Retaining Plate	.20
OS628	Left Extension, Male	.40
OS629	Right Extension, Female	.40
OS638	Bracket for New Style Bolt	.80
OS1561	Valve Seat Cap, including Screen	.40

Parts for 1140 (10) "Old Sol" Bicycle Gas Lamp.

OS74	Carbide Retaining Disc only	$0.20
OS84	Water Filler Cap, with Leather Washer	.20
OS98	Rubber Gasket	.20
OS100	Binding Screw for Head and Generator	.20
OS106	Burner 1/6 ft. Standard Pillar	.40
OS107	Carbide Basket	.60
OS398	Valve Seat, Knurled	.20
OS453	Retainer Ring for Door Glass	.10
OS467	Valve Rod, complete with Thumb and Spring	.80
OS469	Carbide Chamber only	2.00
OS475	New Style Reflector	.40
OS487	Spring Retainer for Filter Plate	.10
OS489	Felt Filter	.10
OS502	Filter Retaining Plate	.20
OS628	Left Extension, Male	.40
OS629	Right Extension, Female	.40
OS638	Bracket for New Style Bolt	.80
OS1561	Valve Seat Cap, including Screen	.40

Parts for No. 1142 (23) "Old Sol" Bicycle Gas Lamp.

OS74	Carbide Retaining Disc only	$0.20
OS84	Water Filler Cap, with Leather Washer	.20
OS94	Spring for Carbide Plate	.10
OS98	Rubber Gasket	.20
OS100	Binding Screw for Head and Generator	.20
OS107	Carbide Basket	.60
OS398	Valve Seat, Knurled	.20
OS453	Retaining Ring for Door Glass	.10
OS454	Red Glass	.20
OS463	Front Door, Glass White	.20
OS467	Valve Rod, complete with Thumb and Spring	.80
OS469	Carbide Chamber only	2.00
OS487	Spring Retainer for Filter Plate	.10
OS489	Felt Filter	.10
OS502	Filter Retaining Plate	.20
OS1561	Valve Seat Cap, Includes Screen	.40

"Old Sol" Lamp Parts—(Continued)

Parts for 1145 (55) "Old Sol" Bicycle Electric Lamp.

No.	Description.	Price Each.
OS103	Door Glass	$0.30
OS170	G. 8—2.8 Volt, 2 C.P. Single Contact Ediswan Base	.44
OS628	Left Extension, Male	.40
OS629	Right Extension, Female	.40
OS1399	Retainer for Door Glass	.30
OS1449	Wiring for Electric Outfit, Includes 1582	1.40
OS1511	Reflector only	1.20
OS1538	Front Fibre Insulator	.10
OS1749	Inside Battery—Connector	.20
OS1574	Battery Case, New Style for Screw	2.50
OS1575	Battery Case Cover, New Style for Screw	1.00
OS1582	Switch Plug	.60
OS1745	Paper Insulator	.10
OS1750	Paper Insulator	.20

Parts for No. 1144 (4) "Old Sol" Bicycle Electric Lamp.

OS628	Left Extension, Male	$0.40
OS629	Right Extension, Female	.40
OS1449	Wiring for Electric Bicycle Outfit, Includes 1582	1.40
OS1469	2.8 Volt, 2 C.P. Double Contact Ediswan Base Bulb	.44
OS1538	Front Fibre Insulator	.10
OS1574	Battery Case, New Style for Screw	2.40
OS1580	Door Glass	.20
OS1582	Switch Plug	.60
OS1584	Retaining Ring	.40
OS1745	Paper Insulator	.10

Parts for No. 1173 (40) "Old Sol" Electric Side Car Lamp.

OS103	Door Glass	$0.30
OS1119	Searchlight Socket, Double Wire, Permanent Plug	.70
OS1168	6 Volt, 2 C.P. Double Contact Ediswan Base	.44
OS1247	Reflector	1.20
OS1399	Retainer for Door Glass	.10

Parts for No. 1186 (5) Generator.

OS4	Valve Rod	$0.80
OS84	Water Filler Cap, with Leather Washer	.30
OS94	Spring for Carbide Plate	.10
OS96	Carbide Chamber, includes Screws only	3.00
OS101	Water Chamber, complete	4.00
OS112	Carbide Basket, with Perforated Tube	.80
OS113	Disc Plate	.50
OS115	Rubber Gasket	.20
OS398	Valve Seat, Knurled	.30
OS487	Spring Retainer for Filter Plate	.10
OS488	Felt Filtering Pad	.10
OS501	Filter Plate	.30
OS1561	Valve Seat Cap, includes Screen	.40

Parts for 1169 (15) "Old Sol" Motorcycle Tail Gas Lamp

OS453	Retaining Ring for Door Glass	$0.30
OS454	Red Glass	.30
OS1155	Burner	.50

Parts for 1171 (64) "Old Sol" Electric Motorcycle Tail Lamp.

OS1154	Retaining Ring for Door Glass	$0.20
OS1153	Red Glass in Frame	.30
OS1056	6 Volt, 2 C.P., Ediswan Base Single Contact Bulb	.44
OS1759	Plug and Bulb Socket	.40
OS1767	Cap for Switch Plug	.20

Lamp Tubing

Rubber Lamp Tubing, for connecting up motorcycle and bicycle lamps with generators.

No. 19251—$\frac{3}{16}$" x $\frac{1}{16}$" wall ..Per 100 ft.

"Old Sol" Burners

No. 1206 (OS106)—1/6 foot, for "Old Sol" Bicycle Carbide (Gas) Lamps .....Each

No. 1207 (OS200)—½ foot, for "Old Sol" Motorcycle Carbide (Gas) Lamps, Nos. 1161 (16), 1164 (17), and 1166 (8)..Each

No. 1208 (OS39)—⅜ foot, for "Old Sol" Motorcycle Carbide (Gas) Lamps, Nos. 1163 (2) and 1168 (13) ....Each

Bray's "Roni" Burner

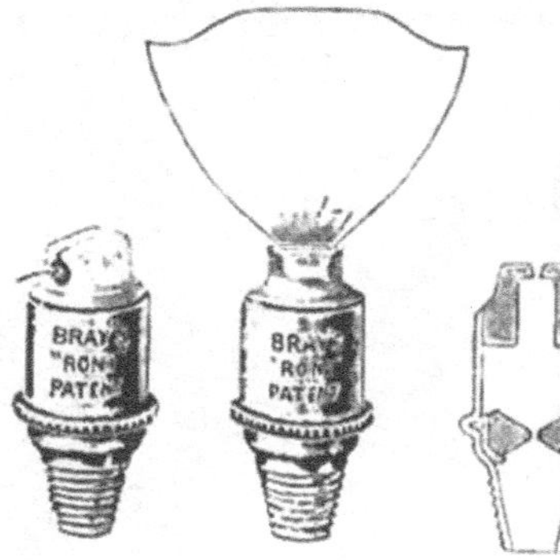

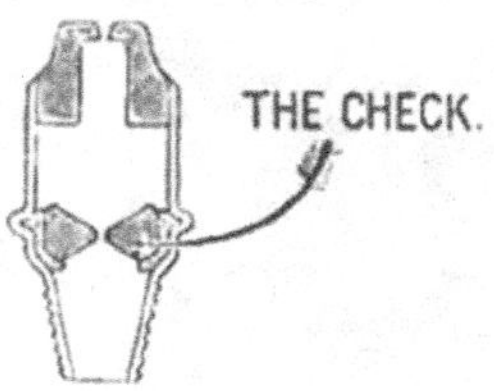

Used on P. & H. Motorcycle Lamps. This atmospheric burner gives a flat flame from a single gas-way. Has no projecting arms to fracture, or jets of flame to get out of alignment; so there is no risk of broken lens or mirrors.

Fitted with check, as illustrated, which prevents flaring when subjected to an excess of pressure.

No. 1204—125 or 127 P. & H. Lamp, ⅜ ft., 10 litres (one dozen in small carton)...Per dozen

No. 1205—For 120 P. & H. Lamp (not listed), ½ ft., 14 litres (one dozen in carton).Per dozen

No. 1200-1.

Motorcycle Lamp Burners

"Alco Hexo" Burner.

Fits them all, and with hexagon base is easiest to instal. Made in ⅜ and ½ ft. gas consumptions.

No. 1200—⅜ ft. (one dozen in box) .... Per dozen

No. 1201—½ ft. (one dozen in box) .... Per dozen

"Alco Cyco" Burner.

A cheaper and very popular style for Motorcycle Lamps. ⅜ and ½ ft. gas consumption. Round knurled base.

No. 1202—⅜ ft. (one dozen in box) ....Per dozen

No. 1203—½ ft. (one dozen in box) ....Per dozen

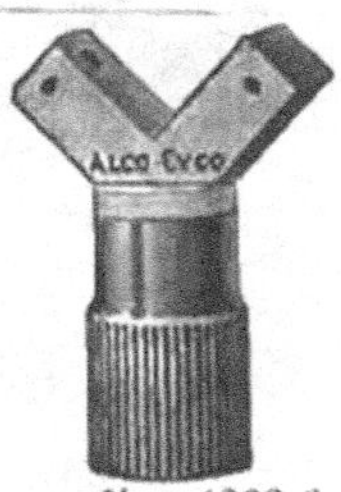

No. 1202-3.

Leggings, Motorcycle

Knee Length.—Well made, from good quality of waterproof canvas. Beaded edge. Fitted with fork fastener. Sizes 15 and 16.

No. 1210 — (Size 15.) Per pair. ....

No. 1211 — (Size 16.) Per pair. ....

Three-quarter length.

Strongly constructed of durable waterproof canvas with beaded edges. Clasp fasteners in place of strap shown in illustration.

No. 1227 — (Size 15.) Per pair. ....

No. 1228 — (Size 16.) Per pair. ....

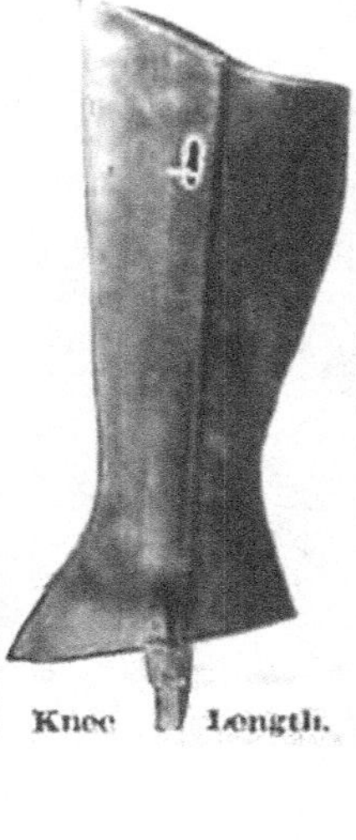

Knee Length.

Three-quarter Length

Locks, Bicycle and Motorcycle

Hex Padlock.—A very popular, fast selling lock, nickel plated, supplied with two keys. Shipping weight 2 lbs. per dozen, packed one dozen locks in box.

No. 1230—(103) . . Per dozen

Mars Padlock.—Another fast selling small, safe lock. Spring opening and self-locking, with two keys. Nickel plated. Shipping weight 2 lbs. per dozen, packed one dozen locks in box.

No. 1230—(100) . . . . Per dozen

Star Padlock.—A neat, well made, spring opening, self-locking padlock with two steel keys for each lock. Size 2 x 2½", nickel plated. Weight 3 lbs. 13 ozs. per dozen. Packed one dozen in box.

No. 1232—(140) . . Per dozen

Assortment (B1).

This assortment consists of one dozen assorted Bicycle Locks and Chains, 4 brass finished and 8 japanned. On handsome lithographed display card. Two keys to each lock.

No. 1218 (B1)—Price per card (one dozen locks and chains) . . . .

No. 1221—Chain only, 9½" long. White metal, double ring (2 dozen in box) . . . . Per dozen

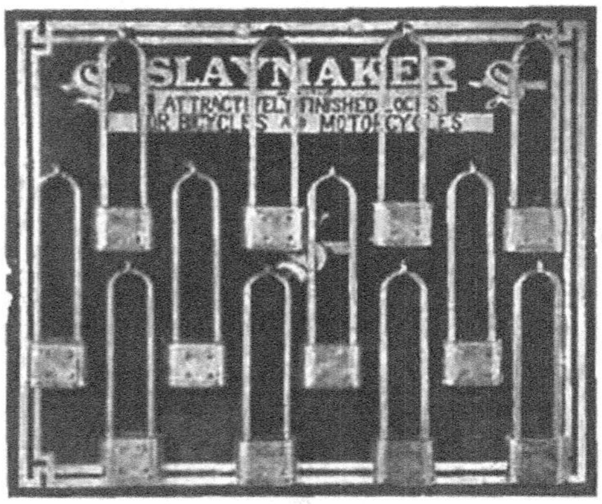

Bicycle and Motorcycle Padlock Assortment displays one dozen locks in assorted finishes. Two each in brass plated, antique coppered, black, blue, green and red enamel finishes.

Locks are 1⅝" across case, and shackle opening is 4¼" x 13/16" inside measurement. Display card is 16½" x 12 ¾", made of extra heavy binder board, with wire easel back and holes for hanging.

Two keys with each lock, keys being on back of card.

No. 1224 (78A) . . . . Per dozen

No. 1215 Bicycle Lock,

Brass Lock, hinged type shackle.

Size 3" over all, Shackle opening ⅞" wide, 2 3/16" deep. Self locking, two keys.

No. 1215 (778)—(One dozen in box) . . . Per doz. . . . .

No. 1222, Bicycle Lock

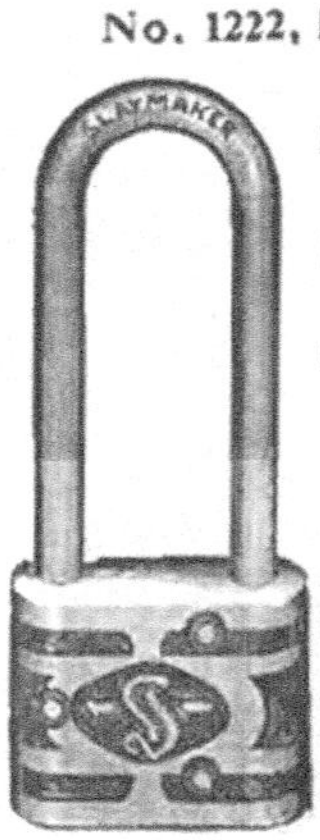

Shell—Polished Brass.

Shackle — Steel Rod, Nickelled.

Interior—Spring Levers.

Key Cylinder—Brass slotted.

Keys — Corrugated and nickelled. 2 to each lock.

No. 1222 (2092)—(6 in box) . . . . . . Each

Black japanned with double ring white metal chain, two keys.

No. 1216 (4026)—(One doz. in carton) . . . . . . Per dozen

MANY a dealer has doubled his Lock sales by merely reminding customers that a bicycle is too valuable and too easily stolen to be left standing without a lock.

Locks (Continued)

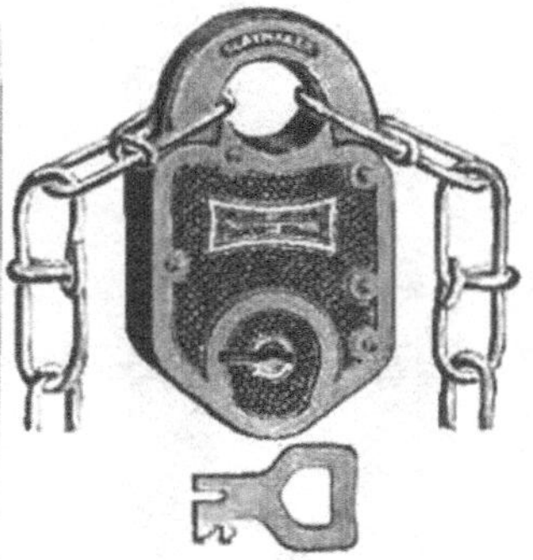

Chain Lock

No. 1219.

Bronze metal case, polished; heavy malleable iron shackle, sherardized (rust-proof), secure brass tumblers, dog and dog detainer. Two heavy flat steel keys each.

Each lock with 18" strong steel wire chain, galvanized.

No. 1219—Price, per dozen

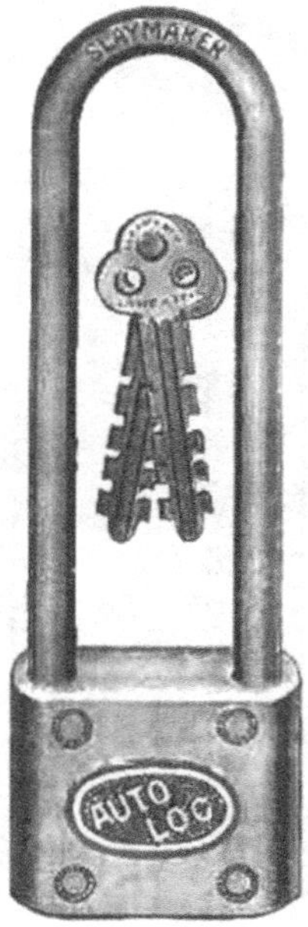

No. 1223—Bicycle and Motorcycle Lock.

Shackle—Hardened steel rod, nickelled, which springs to quarter turn when unlocked. 5" long.

Keys—Two nickelled corrugated steel keys with each lock.

Packed, six locks in carton.

No. 1223 (5-089) .....Each

No. 1220—Motorcycle Lock—Shackle is reversible, adjustable and easily removable from the case. The size of shackle (6¾" long x 1⅛" wide inside measurement) allows the use of this lock on all makes of machines. It is ratcheted to permit adjustment to length of 5⅝".

Brass case 2" across.

Each in a carton with two corrugated steel keys.

(Six cartons packed in a box.)

No. 1220 (1094)—Price, each

Our line of locks will stand any weather and climate, and will not rust or tarnish. Every lock is furnished with two keys.

Lubricant

Chain

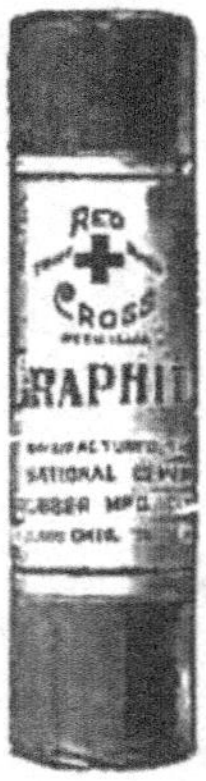

No. 1225.

It is made of the very best material, and is supplied in ⅝" x 2¾" sticks, and 1" x 4" tin box with push bottom.

No. 1225—⅝" x 2¾" Stick (12 in box)....Per doz.

No. 1225—⅝" x 2¾" Stick (12 in box)..Per gross

N. 1226—1" x 4" Tin Box (12 in box)....Per doz.

No. 1226.

Lubricant, Chainolene

The correct lubricant for Roller Type Chains, its distinctive adhesive qualities keeping the rolls revolving continually in a film of oil, thus minimizing the friction. **Not graphite.**

Chainolene.

Ribbon shape opening in tube makes it easy to apply to chain. Economical, because it will not drip or "throw."

No. 368—Collapsible tubes, small size (1 dozen in display box) ....Per dozen

No. 369—Collapsible tubes, large size (1 dozen in display box) ....Per dozen

Nuts

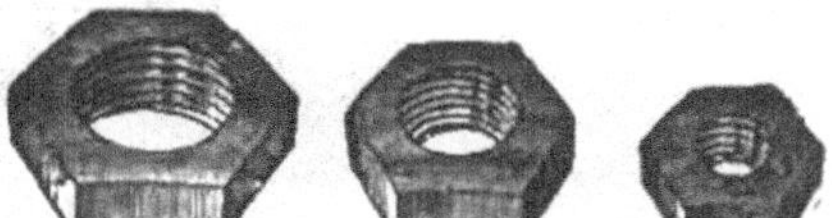

(Hexagon Forged) Nickelled.

Carefully cut and threaded. Hardened to the proper temperature. Well nickelled over copper.

Size.	Thread.
¼	—20, 22, 24, 26, 28, 30, 32 or blank..Per dozen
9/32	—22, 24, 26, 28, 30, 32 or blank......Per dozen
5/16	—20, 22, 24, 26, 28, 30, 32 or blank..Per dozen
⅜	—20, 22, 24, 26, 28, 30, 32 or blank..Per dozen

Nuts, Wing

No. 1300—3/16 Blank, nickelled ....Per 100

No. 1301—¼ Blank, nickelled ....Per 100

No. 1302—For Telescope Pumps. ....Per 100

Oils, Lubricating

"3 in 1" bottle.

Original 3-in-1

Lubricates chain bearings and all points of friction. Never gums. Cleans all parts quickly and thoroughly. This oil is very popular, and, as its name denotes, it is "three in one," being in general use for lubricating, cleansing and polishing.

No. 1353—1 oz. bottles (24 in box) .... Per doz.

No. 1354—3 oz. bottles (12 in box) .... Per doz.

No. 1355—8 oz. bottles (Factory size) (12 in box) ...... Per doz.

"Whiz."

No. 18217—(36 in wood box) In 8 oz. Spout Can. ..... Per doz

(Made in Canada)

C.C.M. Lubricating Oil.

In 2 oz. Bottles.

No. 1358—(12 in carton).. Per doz

Oil
C.C.M. Brake

A specially prepared heavy body oil, for use in the Hercules brake. Use of this oil will give you much better service from your brake.

No. 1383—2 oz. bottles.. .... Per doz.

Hercules Brake Oil.

C.C.M. Burning Oil.

(Made in Canada)

Oil
C.C.M. Burning

A highly refined smokeless oil, for burning in bicycle lamps, motor lamps, carriage lamps, lanterns, etc. Supplied in 8 oz. bottles.

We have a quantity of 8 oz. tins, which will be applied on all orders until stock is exhausted.

No. 1359—(12 in box).. .... Per dozen

Oilers, Bicycle

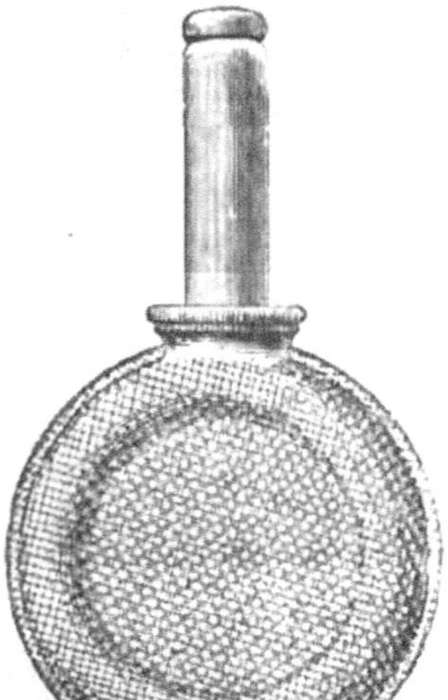
No. 1375.

No. 1375 Popular Oiler, fitted with cap over stem to prevent leaking. Nickelled (½ gross in box). Illustration is actual size. Per doz.

No. 1376—All brass, heavily nickel-plated. Fitted with heavy cap over stem to prevent leaking. Illustration is actual size. This is the oiler supplied in tool bags of C.C.M. Bicycles.

No. 1376 (½ gross in box) .... Per dozen

Oiler
Motorcycle

Motorcycle
Oiler.
No. 1381.

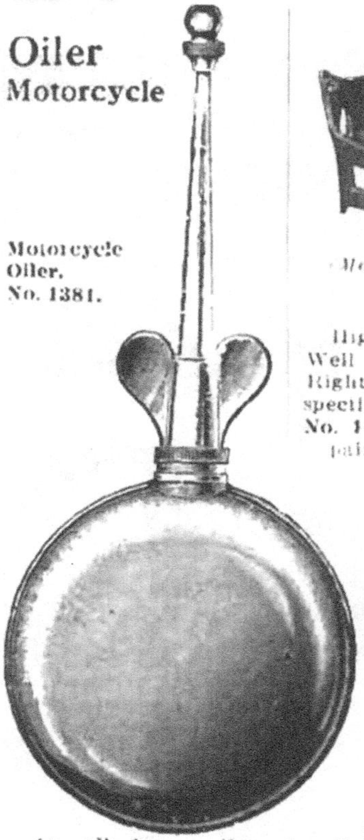

An all brass oiler, heavily nickel-plated. Size 2¾" diameter, length over all 6⅝".
No. 1381—(12 in box) . . . . Per dozen

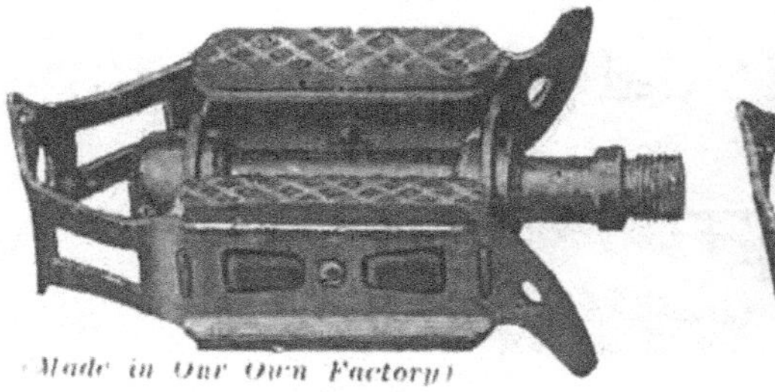

(Made in Our Own Factory)

Hercules Pedal, Men's

High class rubber pedal, very handsome design. Well made, finely finished and easy running. Right and left pedals stamped "R" and "L" respectively.
No. 1401—½" x 20 thread, width 4" (one pair in box) . . . . Per pair

(Made in Our Own Factory)

Hercules Pedal, Ladies

Well made, nicely finished rubber pedal, similar to the Men's Hercules, but with shorter and rounder frame. Right and left pedals stamped "R" and "L" respectively
No. 1400—½" x 20 thread, width 3½" (one pair in box) . . . . Per pair

(Made in Our Own Factory)

Rat Trap Pedal, No. 40

Comparatively new design which is giving splendid satisfaction and promises to be very popular. Accurately made and beautifully finished.
No. 1411—$\frac{9}{16}$" x 20 thread, width 4" (one pair in box) . . . Per pair
No. 1412—½" x 20 . . . . . . Per pair

Parts for Hercules Pedals

(Ladies' and Men's).

No. 1423—Rubbers, Men's . . . . Per Set of 4
No. 1424—Plates, Men's . . . . Dozen
No. 1425—Rubbers, Ladies . . . . Per set of 4
No. 1426—Plates, Ladies' . . . . Dozen
No. 325—Bolts and Nuts . . . . Dozen
No. 1431—Axle, Right . . . . **Each**
No. 1432—Axle, Left . . . . Each
No. 1436—Dust Cap. Each
No. 1433—Cone . . . . Each
No. 1434—Locknut. Dozen
No. 1435—Washer. Dozen

Pedal, Rubber Bicycle

No. 1406—Rubber Bicycle Pedal, ½ x 20 (one pair in box) . . . . Per pair
No. 1408—Rubber Bicycle Pedal, $\frac{9}{16}$ x 20 (one pair in box) . . . . Per pair

Parts for Rubber Pedal

No. 1462 (10)—Rubbers . . . (Grey) . . . . Per set (4)
No. 1464 (1) —Axle, Right . . . . Each
No. 1465 (1) —Axle, Left . . . . Each
No. 1466 (14) Dust Cap . . . . Each
No. 1467 (2) —Cone . . . Each
No. 1468 (4) —Axle Nut . . . . Dozen
No. 1469 (6) —Bolts for Rubber . . . . Each
No. 1470 (7) —Nuts for 1469 . . . . Dozen
No. 1471 (5) Axle Washers . . . . Dozen
No. 1472 (3) —Outside Ball Races . . . . Each
No. 1473 (3A) —Inside Ball
No. 1474 (12)—Plates Doz
No. 1171 (12)—Plates Doz
No. 1478 (8) —Rubber Washers . . . . . . Dozen

Parts for C. C. M. Pedals

No. 1427—Rubbers . . . .
No. 1428—Plates . . . . . Dozen . . . . Per set (4)
No. 325—Bolts and Nuts . . . . Dozen
No. 1431—Axle, Right Each
No. 1432—Axle, Left. Each
No. 1436—Dust Cap. Each
No. 1433—Cone . . . . Each
No. 1434—Locknut . . . . Dozen
No. 1435—Washer . . . . Dozen

"Get Home" Motorcycle Pedal Axle

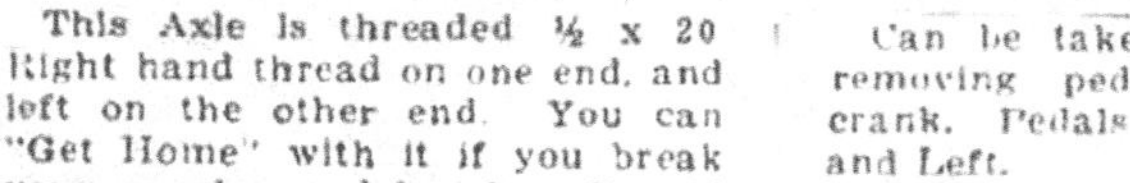

This Axle is threaded ½ x 20 Right hand thread on one end, and left on the other end. You can "Get Home" with it if you break your regular pedal axle. Slotted ends.
No. 1417 . . . . Each

Pedal "Knock Out" Motorcycle

Can be taken off without removing pedal axle from crank. Pedals marked Right and Left.
No. 1418—½" x 20 thread (one pair in box) . . . . . Per pair

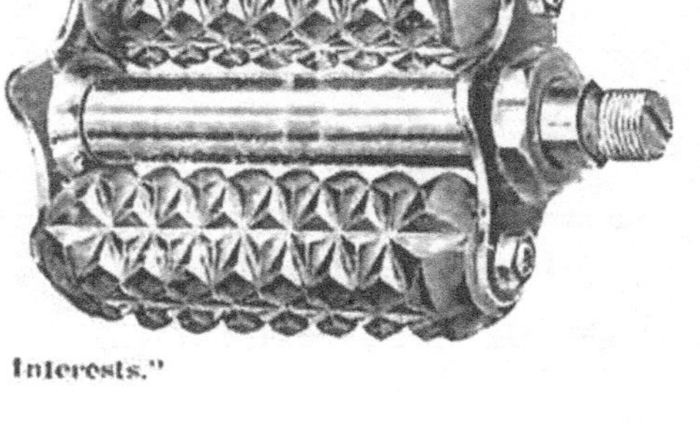

Pedal Wrench, Shop

Specially adapted for attaching and detaching pedals. Opening ⅝", instead of ¾" illustrated. Semi-finished.

No. 2907 (540) .... Dozen

Pennants, Advertising
See page 12.

Pliers
Combination, 6-inch

Manufactured from best tool steel, drop forged and tempered. Black finish, with polished jaws.

No. 15513 .... Per pair

Pliers
Cone, Offset

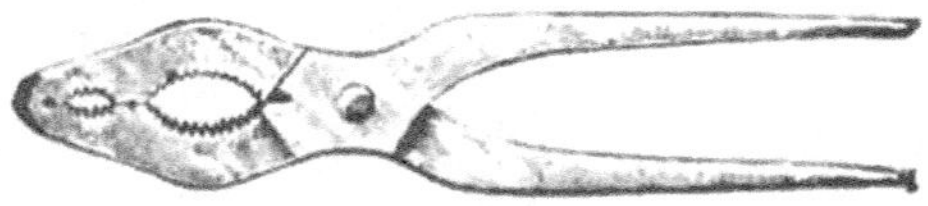

No. 1454.
(Specially hardened in Our Own Factory.)

Drop forged, 8" over all. Has long thin jaws (1/32" thick), offset to grasp large or small lock nuts, cones, etc., where there is only a narrow space to get at them. Has wire cutting attachment and screw driver handle.

No. 1454 .... Each

End Cut Nippers

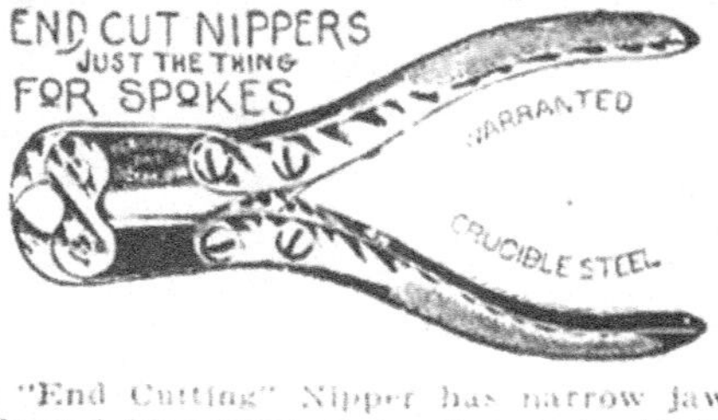

This "End Cutting" Nipper has narrow jaws especially adapted for cutting ends off spokes. In two sizes, 5" and 6".

No. 1452—5" (6 in box) .... Pair
No. 1453—6" (6 in box) .... Pair

Extra jaws for end cutting nippers (In ordering state whether male or female.)

5" .... Each
6" .... Each

For Chain Pliers or Grips
See page 24

Posts, Bicycle Seat

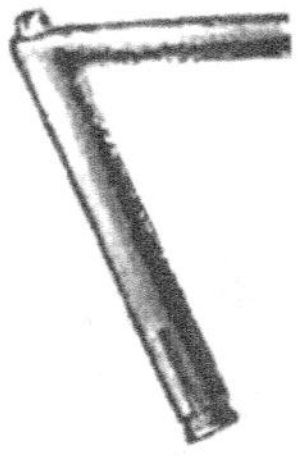

Expander Type.

No. 1510—3 x 5⅞", sizes ¾", 13/16", ⅞", 15/16", 1" ..... Each

No. 1511—3 x 5⅞" (all other diameters) ...... Each

No. 1512 — 4½" x 8½". Sizes ¾", 13/16", ⅞", 15/16", 1". .... Each

No. 1513 — 4½" x 8½". (All other diameters, including 1 1/32", 1 1/16", 1⅛") ..... Each

C.C.M. Type.
Made in Our Own Factory

No. 1514—⅞ x 6" C.C.M. Type. Each
No. 1515—⅞" x 9" C.C.M. Type. Each

NOTE.—The No. 1514 is the Post regularly supplied on C.C.M. high grade bicycles.

Seat Post Clamps
See page 16.

Plugs, Sampson Brass

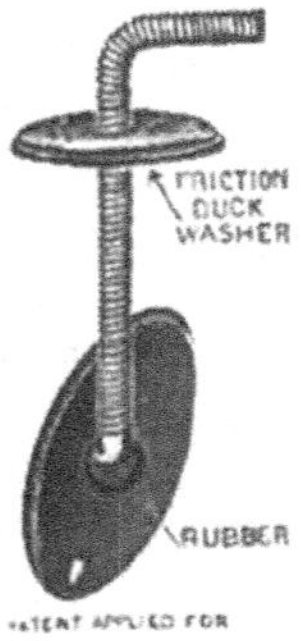

(Made in Canada)

We carry these in three sizes, viz., small, medium and large. 1 gross in box.

No. 1475—Small ...... Per dozen
No. 1476—Medium ... Per dozen
No. 1477—Large ..... Per dozen

Plugs, String

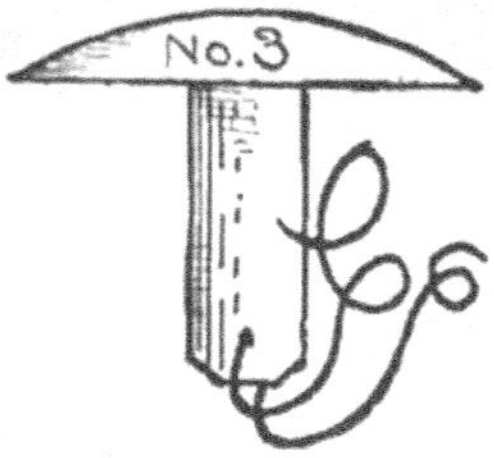

Rubber Plugs — With Strings, for repairing single tube tires, in three sizes: small, medium and large.

No. 1486—Small (100 in box) ..... Per 100
No. 1487—Medium (100 in box) ..... Per 100
No. 1488—Large (100 in box) ..... Per 100

Pumps, Bicycle

Pumps, Hand

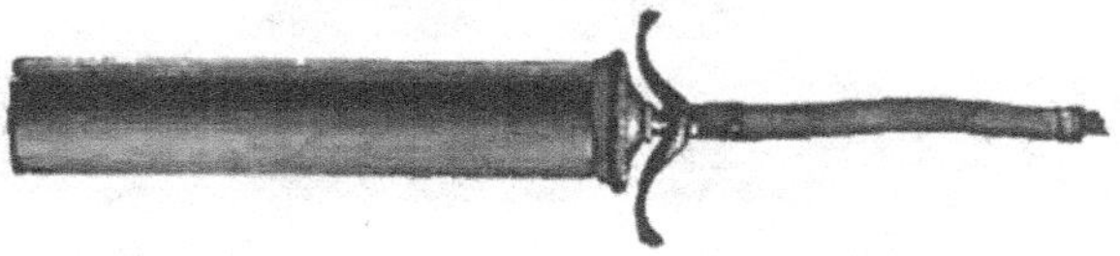

Hand Pump.—5⅜" barrel. Nickel-plated. Supplied with 3" rubber connection, fitting all standard valves.
No. 1520 ....Per Dozen

Pumps, Frame

Austin Bicycle Frame Pump.

Special feature: No tube connection, no leaky joints. End of pump fits over valve as in old "Banner" pump. Size of barrel, 9¼" x ⅞". Length over all, 12½".
No. 1523—(12 in box), complete with one clip... ....Price, each

Hercules, nickelled.—Made from very best quality brass tubing, nickel-plated. Fitted with high grade rubber connection, screw fitting, in end of plunger.
Size of barrel, 9¾" x ⅞"; length over all, 11"
No. 1522—Less Clips (each in carton).Per dozen
No. 1526—With No. 1542 Clips ......Per dozen

Bluemel Black Celluloid.

Very attractive in appearance. Well made and substantial. Solid moulded ends. No parts can come loose with use. Size of barrel, 14¼" x 1⅛". Length over all, 15¼". Complete with rubber connection, with spring-in fitting in end of plunger.
No. 1525—Less clips (36 in box) .....Per dozen
No. 1527—With No. 1540 Clips (36 in box).... ....Per dozen
The 1540 is the English style clip—not listed.

Pumps, Foot

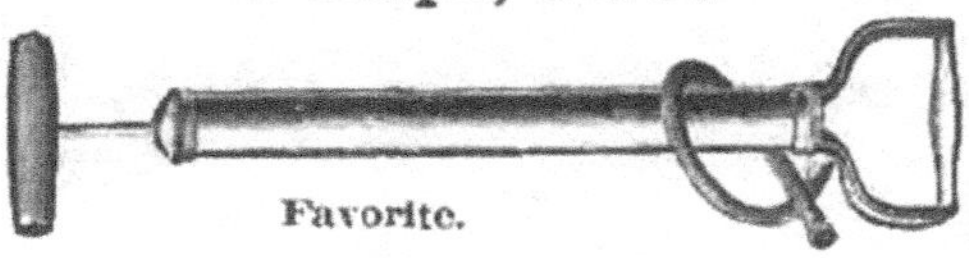

Favorite.

Favorite.—A medium priced Pump, which has given universal satisfaction; 1⅛" x 12", seamless steel barrel, polished. Cap screws off at top. Length over all, 17¾"; weight, 18 oz.
No. 1560 (15)—Favorite (100 in box).Per dozen

"C.C.M."

C.C.M.—1¼" x 12". Length over all, 17¾". Barrel is made from seamless brass tubing, nickel-plated. Cap screws off at top. Weight 19 oz. Cast base. Fitted with 12" rubber pump connection.
No. 1563 (215)—(100 in box) ....Per dozen

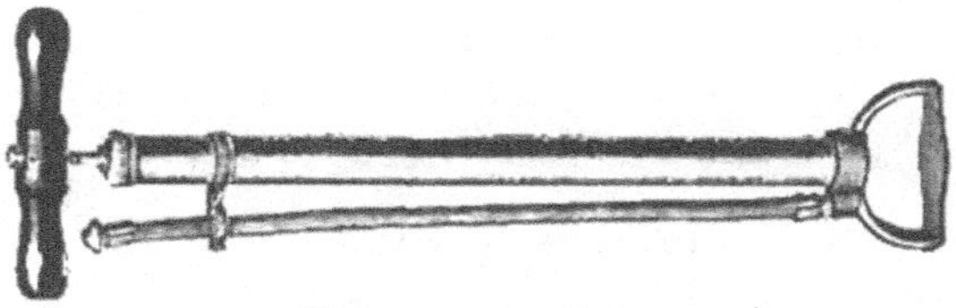

Noera. 1565 (84).

Noera (84)—A good single action Bicycle Tire Pump. Seamless steel barrel, nickel-plated, 1¼" x 13", length over all 23", plated cast base, screw cap. Fitted with 2 foot connection.
No. 1565 (84)—(25 in box) ....Each

Comfort Pump

Two-cylinder double action Pump; cylinder 16" long; height over all, 19¾"; width of handle, 7½".

Steel barrels, cast base. 20" pump connection, with swivel fitting. Exceptional value.

No. 15201 — Less Gauge (50 in box) ......Each

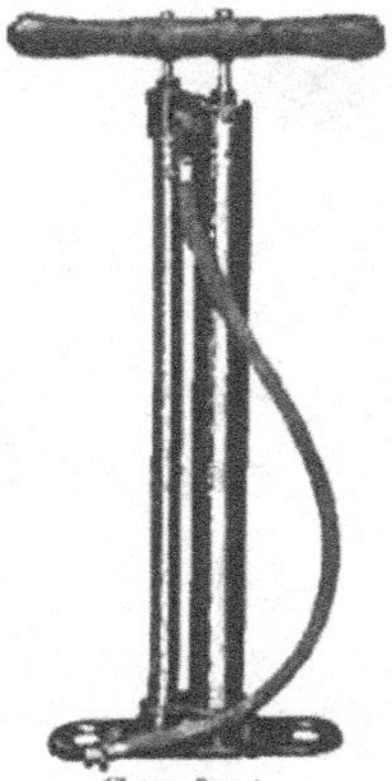

Comfort.

Stands Frame Pump

These will help to sell your pumps. Aluminum finish, 12" at base, 6¼" at apex; height 18½", with standing rack. Takes one dozen pumps.

No. 1550 ....Each

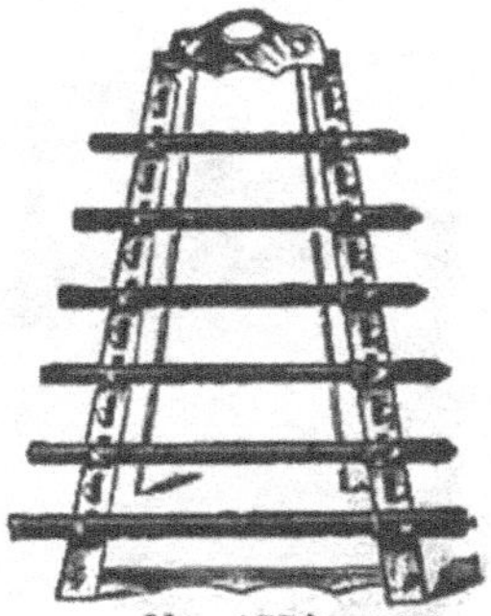

No. 1550.

No. 1542.

Pump Clips, Frame

Suitable for ⅞" pumps. One clip all that is required for holding pump. Fitted with Wing Nut.

No. 1542 ....Per dozen

Pumps (Continued)

Foot Pump, Bridgeport

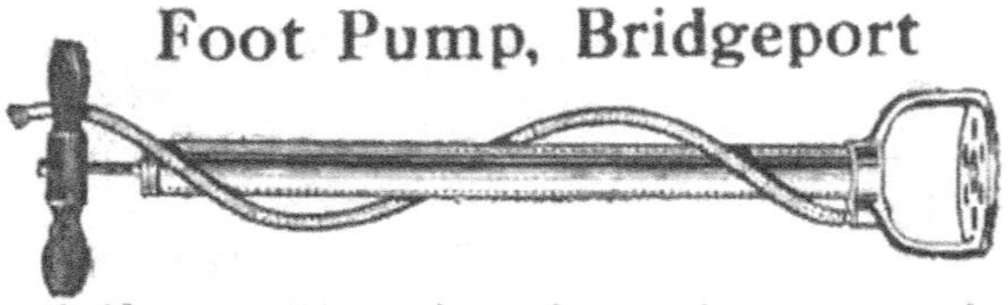

Bridgeport (8)—A large Pump, for store or shop use. Extra heavy seamless brass cylinder, nickel-plated, 20" long x 1½" diameter, length over all 26". Heavy cast base, screw cap. Fitted with 24" connection.
No. 1564 (8)—(25 in box) . . . . Each

Pumps, Motorcycle

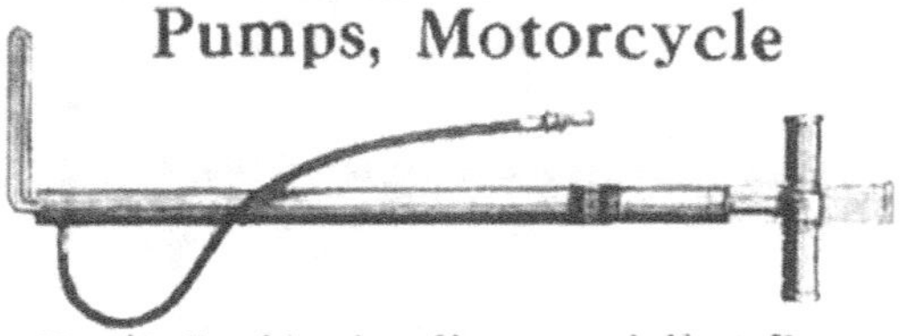

"Excel" Combination Frame and Foot Pump.

Seamless brass barrel, nickel-plated. Size of barrel, 18 x 1"; length over all, 22¾". Fitted with swivel handle, which can be turned lengthwise on end of plunger and fitted over barrel. Foot Bracket also folds against the barrel.

Pump stem fits down inside of plunger with a spring clip. Fits either English or American style valve
No. 1531—Price, with two No. 1542 clips . . . Each
No. 1542—Clips for above . . . . Per dozen

Austin Motorcycle Pump.

Special feature: No tube connection, no leaky joints. End of pump fits over valve as in old "Banner" pump. Size of barrel, 15" x 1. Length over all, 18½". Nickel-plated.
No. 1532—(Each in carton), complete with two clips . . . . Each

Pump Stems

No. 1610—3", for Hand Pumps, plain grey rubber (25 in bundle) . . . . Per dozen
No. 1611—12", for Foot Pumps, plain grey rubber (100 in bundle) . . . . Per dozen
No. 1614—For No. 1564 and 1565 Bridgeport Pumps, two feet long . . . . Each
No. 1615—Universal spring-in style, fits nickel and celluloid English frame pumps. (See note.) (One gross in carton) . . . . Per dozen
No. 1616—For No. 1522 American Nickel Frame Pumps . . . . Per dozen

NOTE.—We cannot obtain a further supply of screw-in style stems for Hercules Nickel and Celluloid Pumps of English manufacture, and have, therefore, arranged for a universal spring-in style stem (No. 1615), which will fit both these pumps.

Our Nickel Pump No. 1522 is now American made, and No. 1616 stem fits it.

Pump Connections

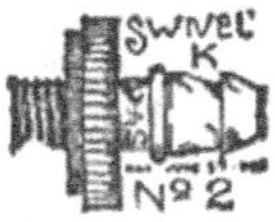

K. Swivel.

Standard equipment on most Foot Pumps, and many Hand Pumps. Handy and easily fitted to rubber hose.
No. 1591—For 3/16" Tubing (25 in box) . . . . Per doz

The **"Keno"**—The threaded part is steel. The stem that goes in hose and the body are one solid piece (25 in box).

No. 1592—Regular size, for 3/16" tubing . . . Each
No. 1593—Large size, for ¼" tubing . . . . Each

Frame Pumps, Motorcycle

No. 1533—Motorcycle Frame Pump, nickel-plated. Size of barrel, 18" x 1"; length over all, 19". Pump connection, with screw fitting in end of plunger.
Complete with two clips . . . . Price, each

No. 1594.

The **"Bull Dog"**—Well made and with a good grip.
No. 1594—(25 in carton) . . . . Each

Pump Washers (Leather)

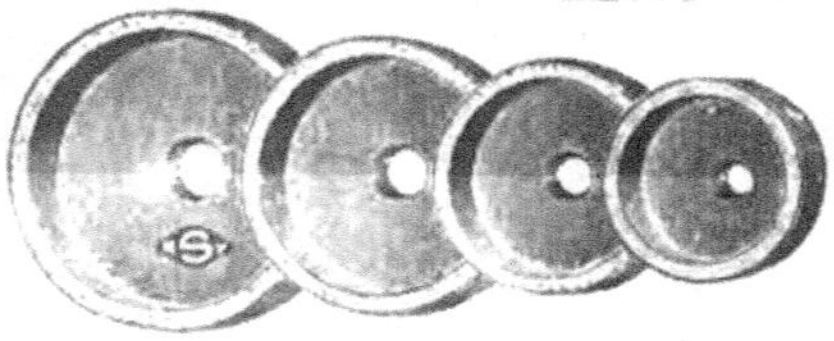

For renewing pump washers on hand, foot and floor pumps. **100 in box.**
No. 22196—⅝" . . . . Per 100
No. 22189—¾". (Fits pumps 1522, 1523) . Per 100
No. 22165—⅞". (Fits pump No. 1520) . Per 100
No. 22166—1" . . . . Per 100
No. 22167—1⅛". (Fits pump No. 1560) . Per 100
No. 22168—1¼". (Fits pumps 1563, 1564). Per 100
No. 22169—1⅜". (Fits pump No. 1564) . Per 100
No. 22170—1½" . . . . Per 100
No. 22171—1¾" . . . . Per 100

Boxed Assortment—For the convenience of the bicycle repair trade, we have prepared an assortment of 72 Pump Washers, comprising the following sizes:—6 ¾", 18 ⅞", 12 1", 6 1⅛", 24 1¼", 6 1½".
No. 1555 . . . . Per box

Pump Connections

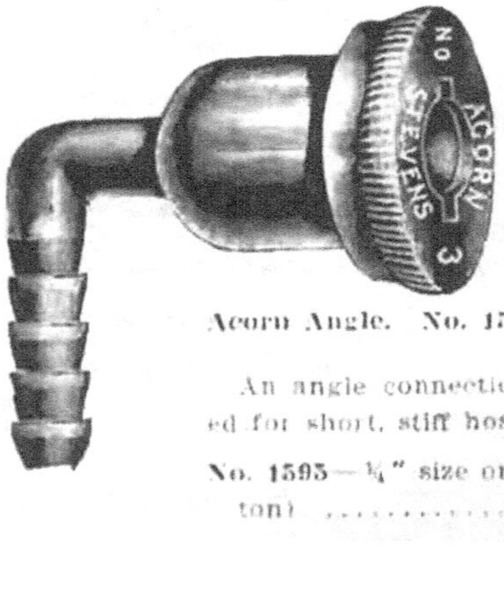

Acorn Angle

Acorn Angle. No. 1595.

An angle connection, especially adapted for short, stiff hose.
No. 1595—¼" size only (25 in carton) . . . . Each

Pump Connections (Continued)

Acorn No. 1596.

Acorn

Our new stock will be fitted with a metal collar inside the chamber, which prevents forcing the connection on the valve so far that it cannot easily be removed.

No. 1596 — 3/16", regular size (50 in package)... .... Each

We cannot obtain ¼" size at present.

Plugs

Benton Mica Motorcycle

No. 1499 Metric.

Made in Two Sizes.

Fully guaranteed. Complete cleaning instructions in each box.

No. 1499 — Metric. Price, each

No. 1500 — A.L.A.M. Price, each

Sectional View.

Thumlock

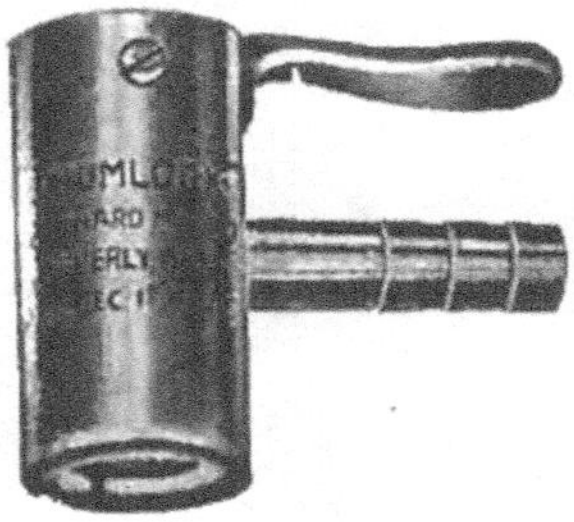

Thumlock Pump Connection.—A new device which grips securely on the valve by pressing a small lever with the thumb. Nothing to get out of order.

No. 1597 ..Each

Reamer

For reaming out valve hole in rims and repair work on wood or metal. Will fit standard stock in brace or breast drill. Tapers from ⅛" to ½". Each in carton

No. 17102 ....Each $1.90

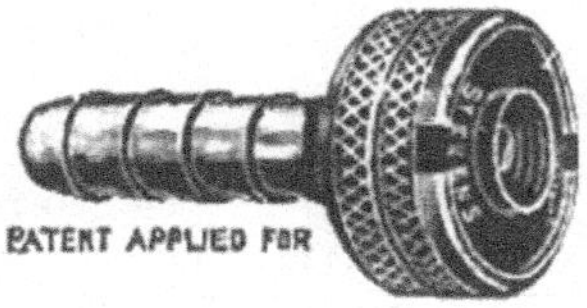

Ideal

For Power Pumps only.

No. 2289—(25 in carton) ..Each

Pump Rubbers

Acorn.

No. 22103 — Acorn (100 in box). Per doz.

No. 1601 — Banner Per doz. ....

Banner.

No. 1538—Fits in Austin Pump ....Per doz

Repair Tags Hercules

(Made in Canada)

Indispensable to every repair man. Each tag is numbered consecutively, and has detachable coupon for customers.

We cannot supply any specified numbers, but will fill orders with consecutive numbers for any quantity.

No. 1655—Per 1,000 (1,000 in pkg.) ..

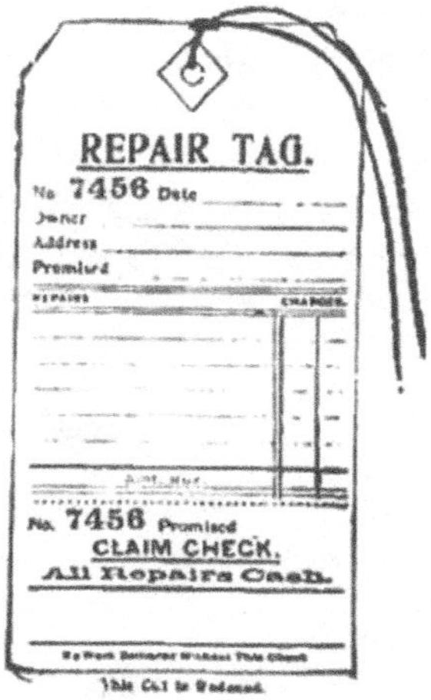

Pump Tubing

(Made in Canada)

We furnish **black** pump hose in three following sizes: ⅛", 3/16", ¼", 2-ply. The ⅛" is for hand or foot pumps, 3/16" for shop or floor pumps, and the ¼" for large automobile power pumps. We also supply a special grade of red tubing, as specified below:

Black.

No. 1620—⅛", 2-ply ....Per foot
No. 1621—3/16", 2-ply ....Per foot
No. 1622—¼", 2-ply ....Per foot

Red.

No. 1630—⅛", 2-ply ....Per foot
No. 1631—3/16", 3-ply ....Per foot
No. 1632—¼", 3-ply ....Per foot

Rivets, Bifurcated

One Rivet Holder will be found in each small box.

Put up in small boxes, 100 assorted. Twelve small boxes in carton.

No. 1680 — Large box (12 small boxes) ... Each.

No. 1681 — Small box (100 assorted Rivets) .... Each..

Rims, Bicycle Wood

Rim Striping Specifications from 1912 to 1918.

All Dunlop Style Single Piece Rims.	No. of Stripe used in **1912.**	No. of Stripe used in **1913.**	No. of Stripe used in **1914.**	No. of Stripe used in **1915.**	No. of Stripe used in **1916.**		No. of Stripe used in **1917-18.**
Cleveland Hygienic	16	22	16	16	16		16
Cleveland Rigid	16	16	16	16	16	Models 455-6	**16
Cleveland Rigid						" 457-8	A18
Cleveland Juvenile (Boys' and Girls')	16	16	16	16	16		A 5
Cleveland Flyer	..	*19	..	..	..		
Brantford Hygienic	6	26	40	40	40		40
Brantford Rigid	20	20	40	40	40	Models 555-6	40
Brantford Rigid						" 557-8	A18
Brantford Juvenile (Boys' and Girls')	20	20	16	16	16		A 5
Massey Hygienic	21	21	25	25	25		25
Massey Rigid	15	25	25	25	25	Models 355-6	**25
Massey Rigid						" 357-8	A18
Massey Juvenile (Boys' and Girls')	15	25	16	16	16		A 5
Massey Grey Wolf	..	*30	..	..	..		
Perfect Hygienic	6	6	16	16	16		48
Perfect Rigid	20	20	16	16	16	Models 655-6	**48
Perfect Rigid						" 657-8	A18
Perfect Juvenile (Boys' and Girls')	20	20	16	16	16		A 5
Gendron Hygienic		26	25	25	25		25
Gendron Rigid	..	20	25	25	25	Models 865-6	**25
Gendron Rigid						" 857-8	A18
Gendron Juvenile (Boys' and Girls')	..	25	16	16	16		A 5
Ivanhoe Hygienic	6	26	16	16	16		48
Ivanhoe Rigid	16	20	16	16	16	Models 755-6	**48
Ivanhoe Rigid						" 757-8	A18
Columbia Hygienic	..	..	..	..	A6 & 8		48
Columbia Rigid	..	..	..	..	A6 & 8	Models 955-6	**48
Columbia Rigid						" 957-8	A18
Model F.	..	..	..	A18	A18		Plain

Motorbike, Model "W," all nameplates. Road Cart Red, with Green Head 36
Green with Gold Head 71

Racer, Model "T," all nameplates. Crescent Laminated Racer style. Section C-6. Plain varnished finish.

***NOTE.—In ordering striped rims for Cleveland Blue Flyer or Massey Grey Wolf, state whether Dunlop or Crescent style required.**

****NOTE.—When high grade Bicycles, irrespective of nameplate, are ordered in any of our stock colors, the following rims will be used:**

Finish No.	Bicycles Enamelled.	No. of Rim equipped.
3	Maroon	40
4	Maroon and Gold Stripe	40
5	French Grey with Green and Red Stripe	30
6	French Grey with Blue Sunburst Head and Blue and Gold Stripe	48
7	French Grey with Red Sunburst Head and Red and Gold Stripe	50
8	Road Cart Red with Green Stripe	36
9	Green with Gold Stripe	71
10	Green with Gold Sunburst Head and Gold Stripe	71
2	Black with Red Sunburst Head	A6
	All "B" Grade Bicycles, men's and ladies', whether black or colored	18
	Juveniles, Green only	71
	Juveniles, Black, Maroon or Grey	A5

Price List

Plain, 24" (36 in crate)	Per pair
Plain, 26" and 28" (36 in crate)	" "
No. 8 Stripe, 28" (36 in crate)	" "
" **16** " 26" and 28" (36 in crate)	" "
" **18** " 28" (36 in crate)	" "
" **25** " 26" and 28" (36 in crate)	" "
" **30** " 28" (36 in crate)	" "
" **36** " 28" " "	" "
" **40** " 28" " "	" "
" **48** " 28" " "	" "
" **50** " 28" " "	" "
" **71** " 26" and 28" (36 in crate)	" "
" **72** " 28" (36 in crate)	" "
" **73** " 28" (36 in crate)	" "
" **A5** " 26" and 28" (36 in crate)	" "
" **A6** " 26" and 28" (36 in crate)	" "
Section C-6, Laminated Rim, Racer Model, (36 in crate)	" "

The above Rim stripings, as specified, will be carried in stock during the coming season, drilled 32 and 36 hole. Any Striping or Drilling different from the above will be special, and unless we happen to have some in stock, will mean a delay of two weeks or longer before delivery can be made.

All Drilled Rims are Fitted with Saw Tooth Washers.

30
8
72
25
73
36
18
PLAIN
A5
40
48
50
71
16
A6

Rims, Bicycle, Wood

(Made in Our Own Factory)

Note.—All drilled wood rims are fitted with saw-tooth washers.

These Rims are manufactured from finest selected hard maple. They are all treated with a special formula waterproof oil, of great penetration, which prevents splitting at the spoke holes, and greatly increases the life of the rim. Finished with highest grade of varnish, and highly polished.

Fully guaranteed as to quality, material and workmanship.

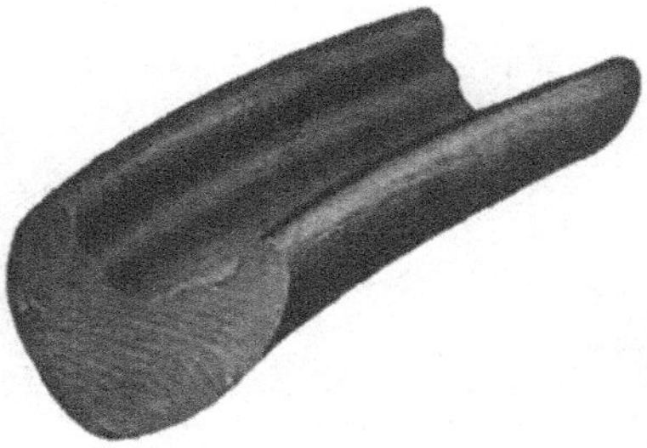

"Dunlop" Single Piece.

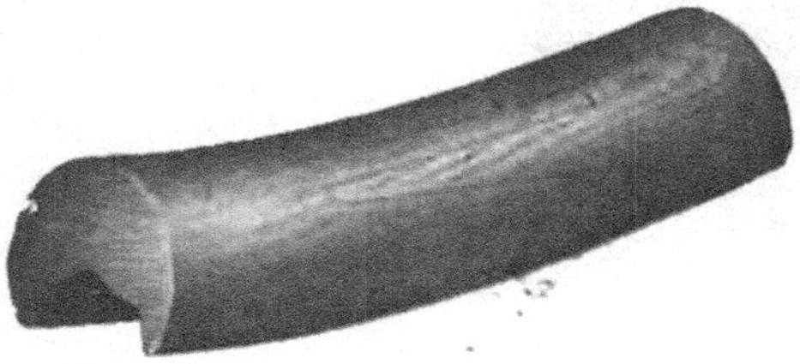

Single Piece, showing Interlocking Joint.

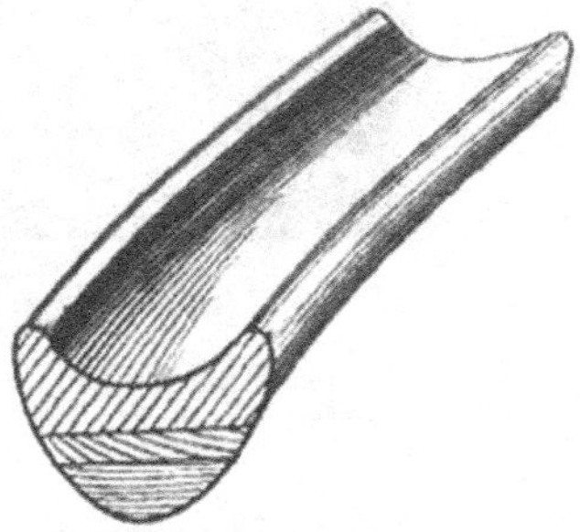

"Crescent" Laminated.

Single Piece—Dunlop, for 24" tires (36 in crate)..Per pair

Single Piece—Dunlop, for 26" and 28" tires (36 in crate) ....Per pair

Crescent Laminated—For 28" tires only (36 in crate).... ....Per pair

Note.—All drilled wood rims are fitted with saw-tooth washers.

For Built-Up Wheels

See page 70.

Rims, Seconds

Rims—Seconds. These are Rims rejected by our inspectors because of imperfect coloring in the natural wood, or a knot or some other slight imperfection which does not impair the serviceable qualities of the Rim in the least.

These Rims are fitted with saw-tooth washers and crated. They are marked "Star." All orders for these are subject to previous sale.

Star—Drilled 32 and 36 hole (36 in crate). Per pair

Rims, Motorcycle, Steel

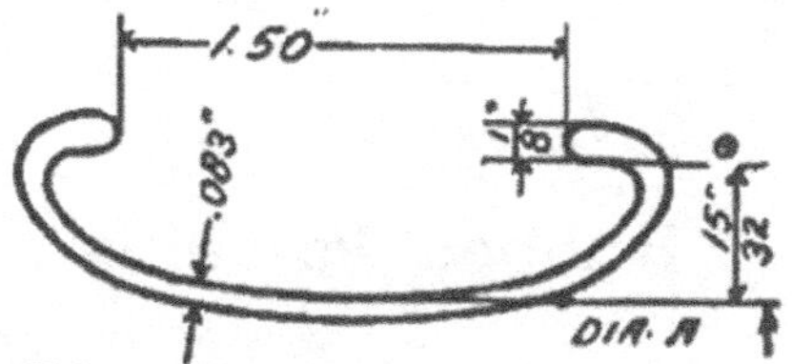

BB.—Takes place of old 2½" Rim, and accommodates 2¼" and 2½" x 28" Tires.

Supplied in rough coppered finish only. Drilled 36 or 40 holes, or undrilled.

Each ....

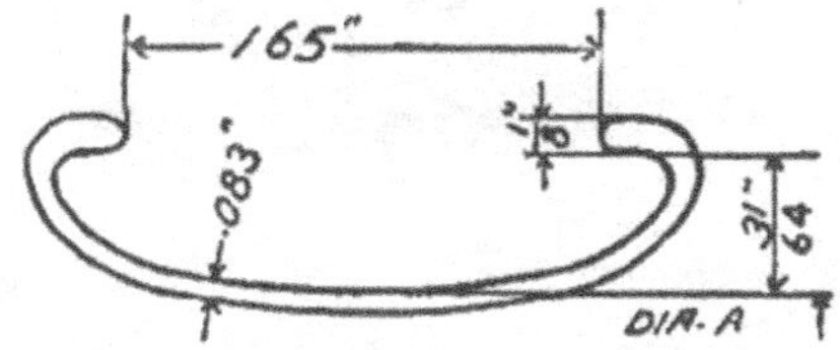

CC.—Made for heavy machines. Will accommodate 2¾" and 3" x 28" Tires ....Each

Supplied in rough coppered finish only. Drilled 36 or 40 holes, or undrilled.

Rims, Steel

Cannot Supply

Owing to the shortage of Steel Rims, we have been obliged to discontinue listing this line for the present.

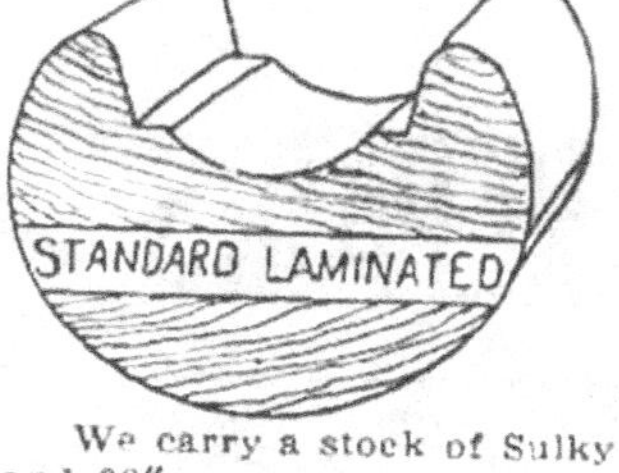

Rims, Sulky or Tandem

Made in Our Own Factory

We carry a stock of Sulky Rims in Dunlop style, 26" and 28".

(18 pair in crate) ....Per pair

Rubber Patches, Circular

Assorted sizes, 100 in box. *(Made in Canada)*

Assorted sizes $1\frac{1}{8}$" to $1\frac{1}{2}$". Very handy for repairing punctures in inner tubes. Packed 100 in box.
No. 1665—Per box of 100 ....

Rubber Patches

M. & M. Sticktite Patch

Sticktite Gasoline Patches solve the problem where an extra quick repair is desired in an emergency.

No. 15048—Quick Emergency Patch, in tin box, containing 10 assorted patches, $1\frac{1}{4}$" to 2" diameter, with strip of sandpaper for cleaning tube, and complete instructions.
Per box ....

Rubber, Patching

Made in Canada

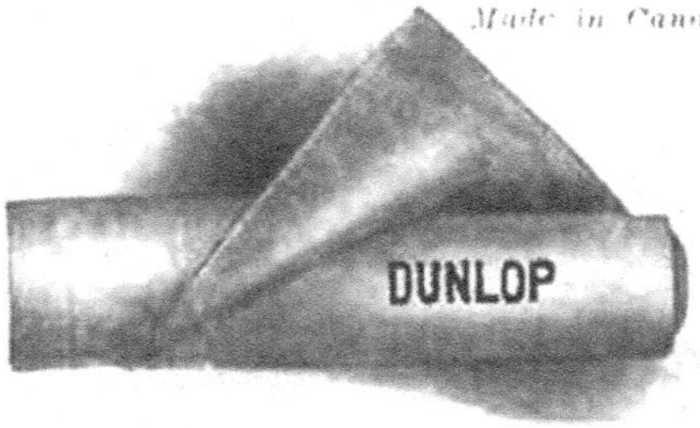

High Grade Rubber, good heavy weight, supplied in 2-oz. rolls, $2\frac{1}{2}$" wide, about 6 ft. long. Specially adapted for repairing tears or punctures in inner tubes. Eight rolls to box.
No. 1660—Per roll ....

Signs, Enamelled

Blue Letters on White Background.—Supplied lettered on both sides, to project in front of shop.
No. 1905 .... Each

Screws

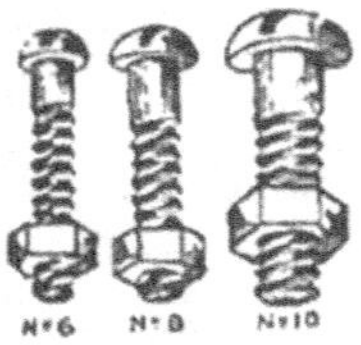

Round head, nickelled. Used on toe clips, bells, mud and chain guard braces, etc. Nickel dipped, cold-pressed hexagon nuts are supplied to fit the screws, if desired. They are supplied in three different sizes, one gross in box, as follows:

Screws Only—No. 6 Wire, $\frac{3}{8}$", $\frac{1}{2}$", $\frac{5}{8}$" long, 32 thread .... Per gross

Screws Only—No. 6 Wire, $\frac{3}{4}$", 1" long, 32 thread .... Per gross

Screws Only—No. 8 Wire, $\frac{3}{8}$", $\frac{1}{2}$" long, 32 thread .... Per gross

Screws Only—No. 8 Wire, $\frac{5}{8}$", $\frac{3}{4}$" long, 32 thread .... Per gross

Screws Only—No. 8 Wire, 1", $1\frac{1}{4}$" long, 32 thread .... Per gross

Screws Only—No. 10 Wire, $\frac{3}{8}$", $\frac{1}{2}$" long, 32 thread .... Per gross

Screws Only—No. 10 Wire, $\frac{5}{8}$", $\frac{3}{4}$" long, 32 thread .... Per gross

Screws Only—No. 10 Wire, 1", $1\frac{1}{4}$" long, 32 thread .... Per gross

Screws Only—No. 10 Wire, $1\frac{1}{2}$" long, 32 thread. .... Per gross

Screw Nuts for above .... Per gross

Screws and Nuts, assorted (1 gross each in box) .... Per box

Shine-M-Up

Hub Cleaners

Put up on display cards holding 6 fronts and 6 rears. Six cards in box.

No. 1900—Per card (1 doz.) ....

Steel Wool

For rubbing down bicycle frames. Supplied in one pound packages in following grades—fine, medium and coarse.

No. 1990—Fine, 1 lb. in package .... Per lb.

No. 1991—Medium, 1 lb. in package .... Per lb.

No. 1992—Coarse, 1lb. in package .... Per lb.

Saddles

"Comfort" and "Hercules"

"Comfort" and "Hercules" Saddles are made from finest selected leather. The springs are manufactured from the best cold-drawn steel wire, specially prepared for that purpose. The binding clips are made true to gauge, ensuring a firm grip on the seat post. These saddles are of first-class finish in every respect, and handsome in appearance.

Comfort (17).

Standard Roadster. Size 10½" x 8" x 4". The three coil springs and the truss are black enamelled; the clamp nickel-plated. Optional equipment on C.C.M. Grade "B" Men's Bicycles.

No. 1817 ....Each

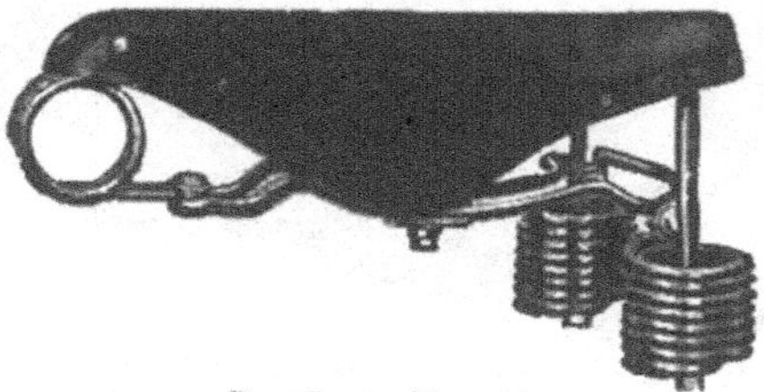

Comfort. No. 42.

10¼" long x 7⅝" wide. Truss and spring black enamelled. American style clamp, nickel-plated. Standard equipment on C.C.M. Grade "B" Men's Bicycles.

Reinforced Seat prevents sagging.

No. 1842 ....Each

Comfort, Ladies' (24).

Size 8½" x 7½" x 4". Truss and coil springs black enamelled. American style clamp, nickel-plated. Standard equipment on C.C.M. Grade "B" Ladies' Bicycles.

No. 1824 ....Each

Comfort, Ladies' (34).

Comfort. Ladies' (34)—Size 10" x 8½" x 4", nickel-plated springs and fittings. Embossed top. Standard equipment on C.C.M. Ladies' Grade "A" Bicycles

No. 1834 ....Each

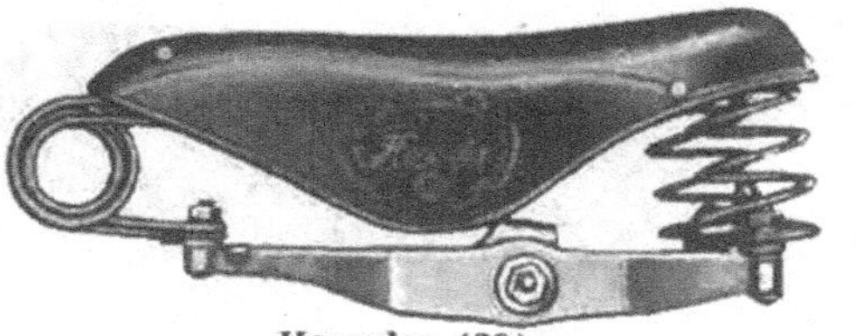

Hercules (39).

Size 10½" x 8⅜" x 4". Heavy rear wire coil springs, double coil steel nose springs, steel girder base. Black enamelled springs and girder, nickel-plated clamp. Very superior leather top, blocked to shape. Well suited for heavy riders. Standard equipment on C.C.M. Grade "A" Men's Bicycles.

No. 1839 ....Each

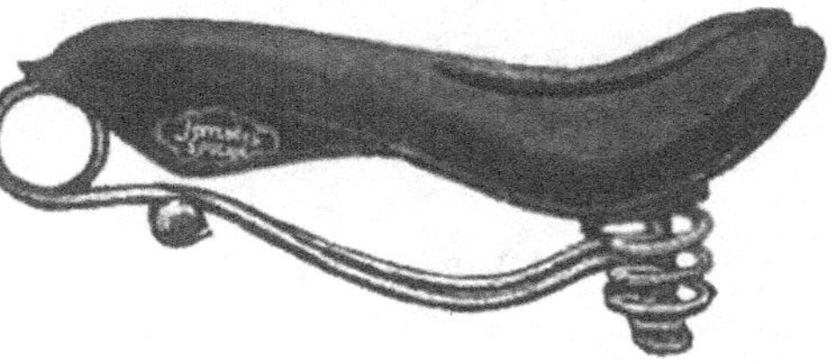

Hercules Special.

Hercules Special Saddle. Handsome, well finished standard road saddle. Length 9", width 7¼". Fitted with Universal spring and special hinge joints on front; double coil torsion spring.

No. 1836 ....Each

Hercules (No. 41). Track Racing Saddle.

Size 11¼" x 6¾" x 2¼". Weight 1 pound 9 oz. Nickelled springs of cold drawn steel wire. A popular saddle for track racers; extra long, with back narrow and well rounded. Standard equipment on C.C.M. Racer Bicycles.

No. 1841 ....Each

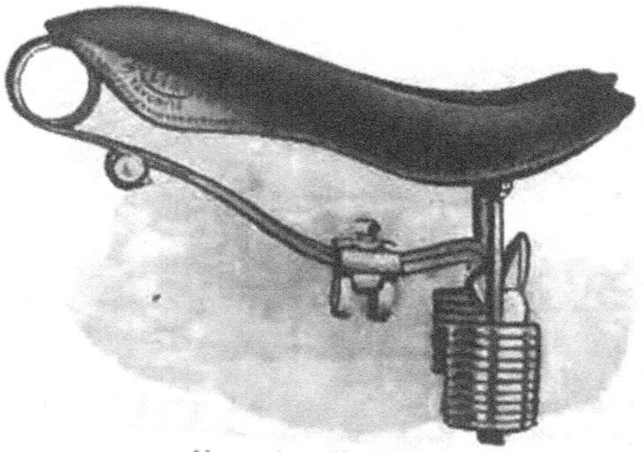

Hercules Favorite.

Hard top saddle, 10⅜" long x 7½" wide. Truss and spring black enamelled. American style clamp, nickel-plated.

No. 1810 ....Each

NOTE.—When ordering parts always send samples.

Saddles (Continued)

Nemo, Men's.

No. 1800—Men's Nemo. Size 10½" x 8" x 3". Black enamelled truss, nickel-plated clamp and nose fitting. Standard equipment on C.C.M. Juveniles, Boys' models ....Each

Nemo, Ladies'.

No. 1801—Ladies' Nemo. Slightly shorter than Men's Nemo; embossed top. Standard equipment on C.C.M. Juveniles, Girls' models. Each

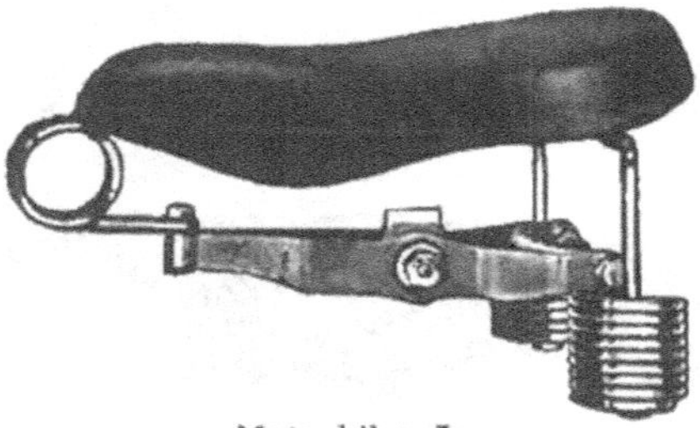

Motorbike Jr.

Motorbike Jr.—A motorcycle type saddle, with padded top and patent compound springs, built to last indefinitely. Size of top 11" x 9". Weight 4 lbs. 10 oz. Metal parts nicely black enamelled. Standard equipment on C.C.M. Motorbike Bicycle.

No. 1844 ....Each

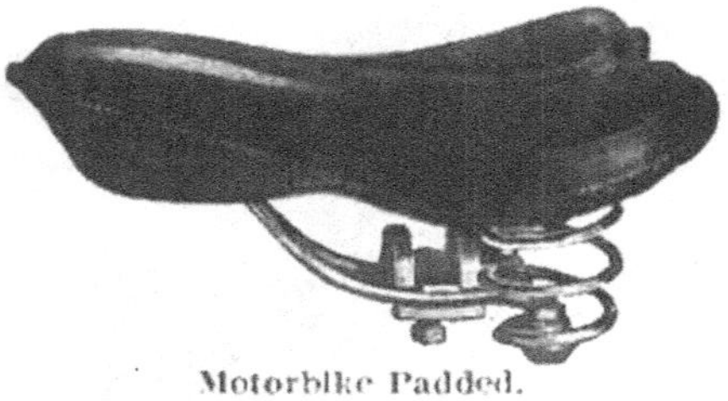

Motorbike Padded.

D-2 Padded—A light, durable, good looking, comfortable saddle. Top consists of five-ply maple and basswood veneer, light and strong, to which is cemented several layers of high grade felt, the whole then covered with grain leather. Fitted with patent compound springs. Size of top 9¼" x 7½". Weight 1 lb. 7 oz.

No. 1845 ....Each

No. 1843. Zenith (3).

Size 10⅝" x 8½" x 4". A heavy, durable saddle, with very serviceable truss and rear springs. Black enamelled girder base.

No. 1843—Zenith (3) ....Each

Motorcycle Saddles

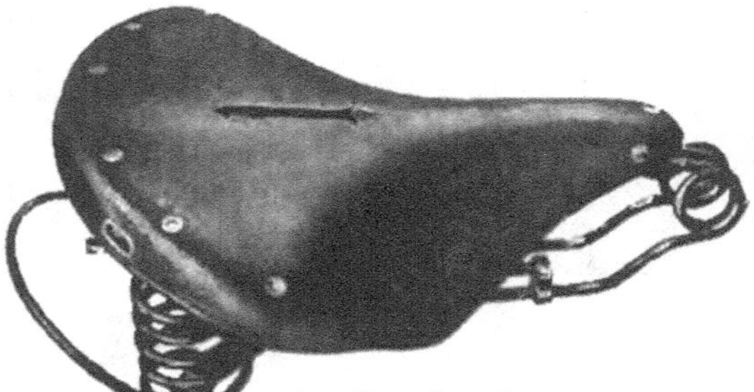

La Grande (30).

Fits low on machine. Size 14" x 11¾". Weight 4 lbs. 9 oz. Highest grade oak tanned leather top, reinforced. Truss black enamelled.

No. 1830 ....Each

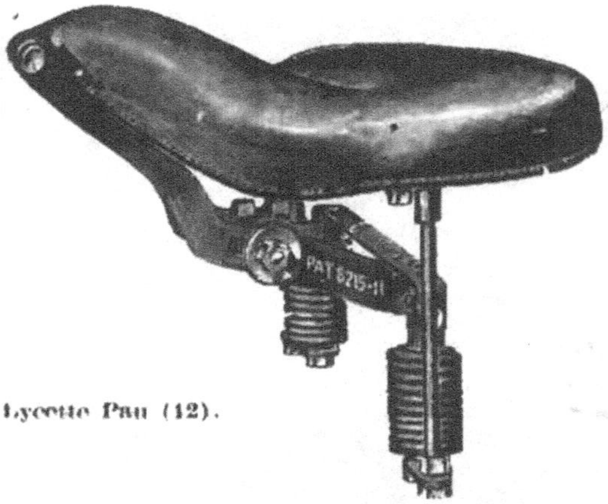

Lycette Pan (12).

A comfortable saddle, made of the best material. Width 13", length 12¾". Base of saddle is of steel, properly shaped to fit, nicely padded, covered with high quality leather. Truss is heavy gauge metal, black enamel finish, with extended Torsion spring, very resilient, under rear of saddle.

No. 1812 ....Each

NOTE.—When ordering parts always send sample.

Prices Subject to Change Without Notice.

Saddle Parts

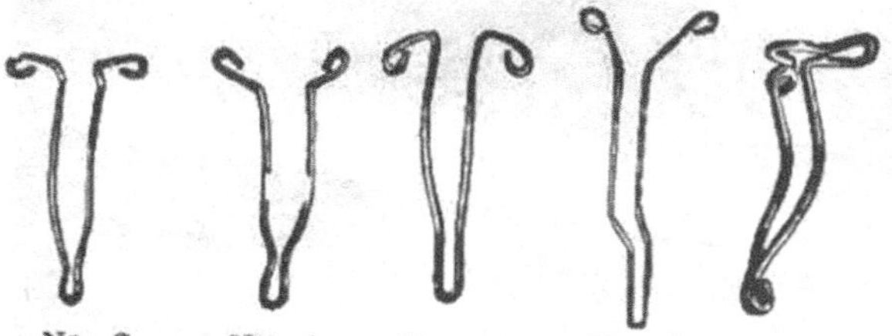

No. 2. No. 4. No. 00 No. 11 Hercules Special.

Saddle Trusses

No. 5.

No. 2—Truss for American Saddle, with two rear coil springs. .... Each

No. 4—Truss for American Saddle, without coil springs.... .... Each

No. 00—Truss for American Saddles, without coil springs, and bolted on saddle base ......Each

No. 11—Truss for Saddle, with three coil springs. .... Each

No. 5—Truss for American Saddles (comprising two coil springs), bolted on saddle base..Each

Hercules Special ....

Saddle Springs

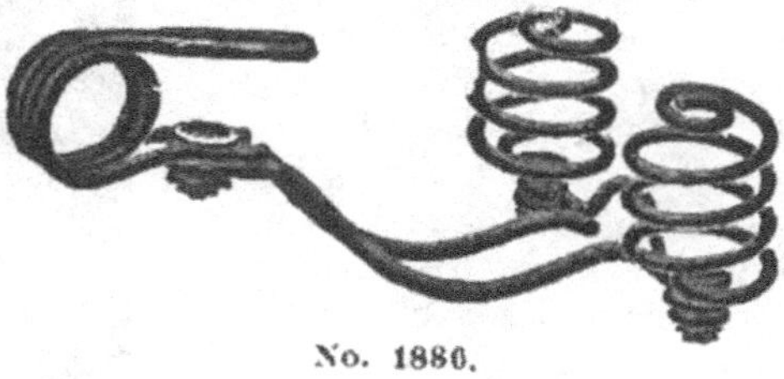

No. 1886.

Troxel Combination Saddle Springs.

An easy saddle spring. Adjustable to any saddle.
No. 1880—Complete ....Each

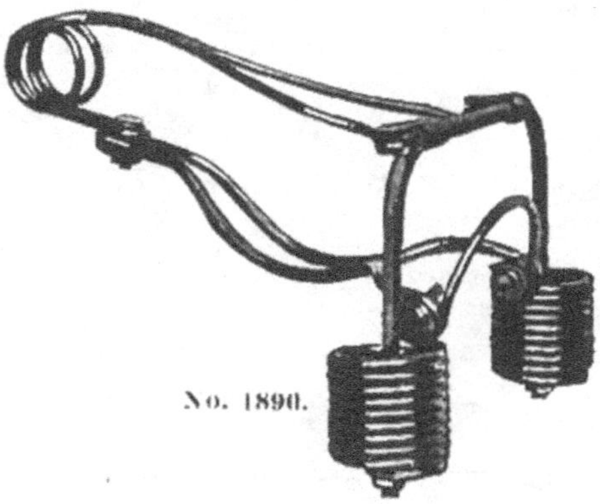

No. 1890.

Perfection Springs.

All assembled and nickelled. A very easy spring.
No. 1890—Each ....

Always Send Sample of Part Required

Saddle Clamps

American—This is the standard Clamp for American styles of saddle springs.

No. 1875—Clamp, complete. Each

Clamp Parts.

Clip only ..Each
Clip Binder only. .... Each
Clip Binder, Set Screw only .. .. Each

Clamp, Chalfant

A Clamp for attaching to top bar of frame for purpose of lowering saddle by bringing it closer to the frame than the seat post would allow. No seat post is necessary when this clamp is used.

No. 574—1" size ....Each
No. 575—1⅛" size ....Each

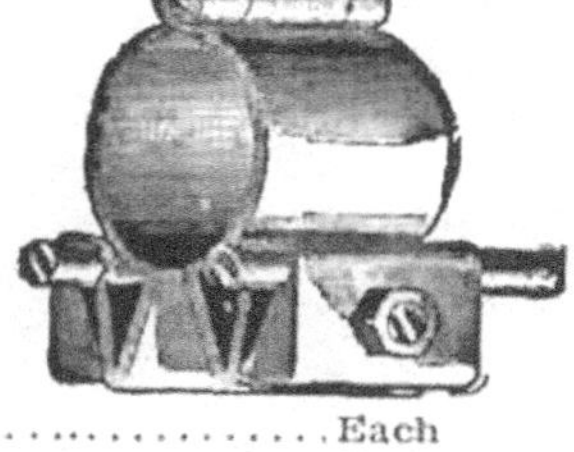

ORDERING SADDLE PARTS

Always send in a sample of the part required when in need of parts for the saddles equipped on our bicycles, or sold by us in the Accessories Department.

This is absolutely essential for the correct filling of your order.

The reason for this is that during the last two years we have been obliged, on account of war conditions, to purchase our saddles from a number of different sources. While we are retaining our own saddle model numbers, naturally the saddles supplied by different makers do not have identically the same parts.

For example, the No. 39 Saddle has been made by three different manufacturers, and each maker's parts are slightly different from those of the others, hence we must have samples.

Spokes and Nipples, Bicycle, "Hercules"

Tensile strength and extra quality finish are outstanding features of Hercules spokes. They are manufactured from the very best grade of piano wire in 15-17 gauge and are tested to a breaking strain of 800 pounds.

Hercules spokes are heavily nickel-plated on burnished copper, and are practically rust-proof. They will retain their high polish and wear for years. They are supplied in neat cardboard boxes, packed 150 spokes and nipples per box, and are furnished in all standard lengths.

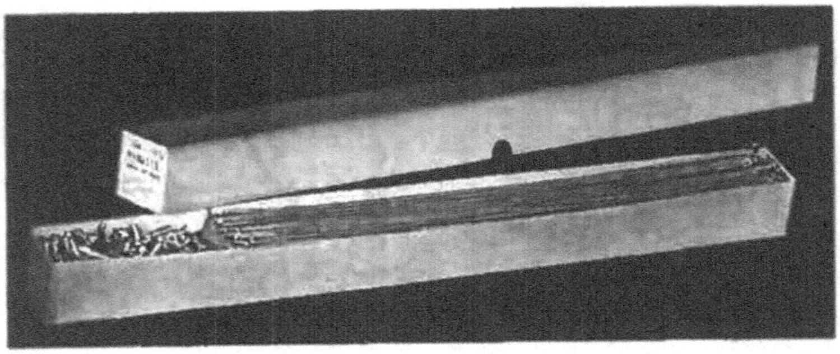

(Made in Our Own Factory)

Notice the quality of the plating on Hercules Spokes.

No. 1920—Swaged—Spokes and Nipples, complete (150 in box) ....Per 1,000

Spokes only ....Per 1,000

Spoke Nipples only....Per 1,000

Note.—Swaged Spokes are not made shorter than 11⅝".

No. 1925—Unswaged—Spokes and Nipples, complete (150 in box) ....Per 1,000

Spokes only ....Per 1,000

Spoke Nipples only...Per 1,000

Note.—Lengths of spokes generally used in bicycle wheels are as follows:

For 24" front wheels.................. 9⅞ and 10"
For 24" rear wheels..................10"
For 26" front wheels..................10¾ and 10⅞"
For 26" rear wheels..................10⅝"
For 28" front wheels..................11¾ and 11⅞"
For 28" rear wheels..................11⅝"

Note

When stringing 36 hole rims, use ⅛" shorter spokes than in 32 hole rims. Example: 28" front wheel, 32 hole rim, use 11⅞" spokes; 28" front wheel, 36 hole rim, use 11¾" spokes.

Spokes and Nipples Sulky

No. 1940—Bent Head Spokes and Nipples, 12-14 gauge, 10¾", 11¾", 11⅞"—Per 100

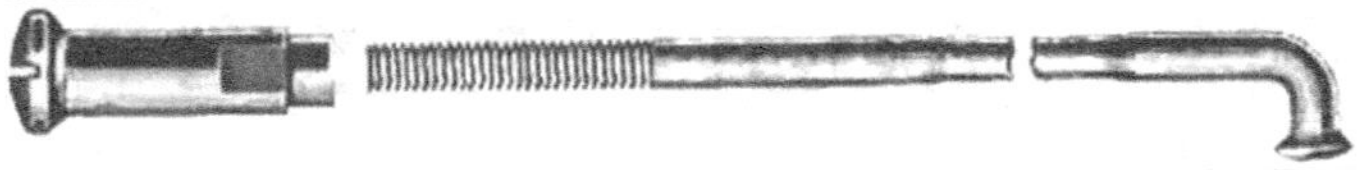

Spokes, Bike Wagon

We supply these in the following gauges and lengths only.

11 gauge, bent head, 14", 15". 11 gauge straight head, 15", 16". 9 gauge, bent head, 14½", 15½", 16⅝".

No. 1930—11 gauge, straight head ....Per 100

No. 1931—11 gauge, bent head ....Per 100

No. 1932—9 gauge, bent head ....Per 100

Skate Grinding Holder

See page 32.

Spokes and Nipples, Motorcycle

No. 1950—Size 135-110 (10-12), all lengths ......Per 100 **$7.80**

Following is a list of lengths used by various manufacturers, practically all of whom adopted the Diamond E. Spokes in 135-110 size as standard equipment on their 1915, 1916 and 1917 models:

10⅛" for 1915 Dayton Rear.
10⅜" for 1915 Dayton Front.
10 5/16" for 1914-5 Pope.
10 7/16" for 1914-5 Pope. 1915 Henderson 2-speed Rear.
10 9/16" for 1914 Harley-Davidson. 1915 Emblem. 1915 Excelsior. 1915 Yale
10⅝" for 1913 Harley-Davidson. 1915 Emblem Front.
10¾" for 1914-5 Indian Front. 1915 Henderson.
10⅞" for 1913-5 Indian Rear (7 h.p. model). 1912-3 Harley-Davidson. 1915 Emblem Rear.
11 1/16" for 1912-13 Pope.
11⅛" for 1912 Excelsior.
11 3/16" for 1912 Pope.
11⅜" for 1912-15 Indian. 1912 Triumph. 1915 Harley-Davidson (a few).
11 7/16" for 1913 Excelsior

For 1916 Models.

Maker.	Spoke Length.
Excelsior	10 9/16"
Hendee—	
Model F-G	10¾" front
	10⅞" rear
5½ H.P.	11⅜"
Model K	10½"
Side Car	10¾"
Harley-Davidson	10 9/16" front
	10 9/16" rear
Henderson	10 9/16" front
	10⅝" rear
Miami	10⅞"
	11⅜"
Reading Standard	10⅝"

Stands, Bicycle

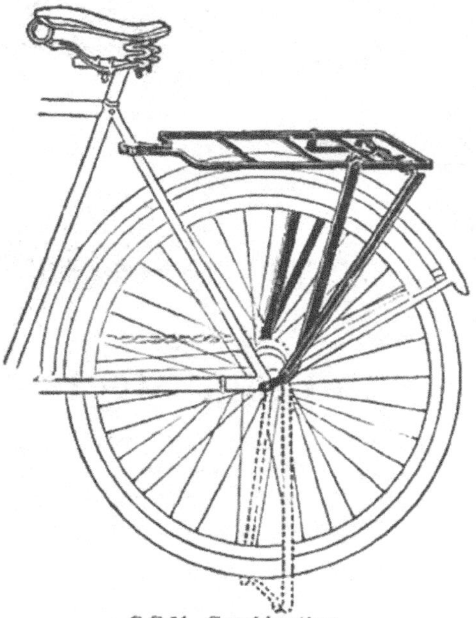

C.C.M. Combination.
(Made in Canada)

No. 412—C.C.M. Combination Rear Carrier and Stand (6 in package) ....
No. 571—Grey Enamel ....Each
No. 572—Maroon Enamel ....Each

For further description, see page 21.

Perfection Floor Stand.

A valuable fixture in any store for properly displaying a bicycle. It is a silent salesman.

No. 1971 ....Per pair

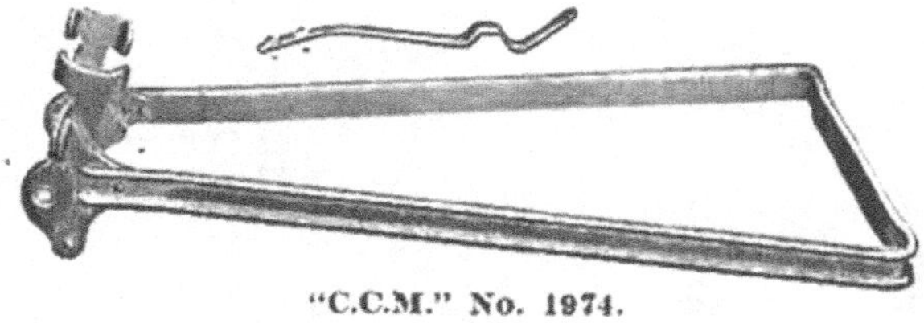

"C.C.M." No. 1974.
(Made in Canada)

Is attached to rear axle and can be used as a shop display stand, or for bicycle delivery work. Made of channel steel, enamelled black. Weight 31 ounces.

No. 1974 ....Each

Myers' Open.

Myers' Floor Stand— Excellent to use in showing off wheels. Adjusted to any wheel. Finished in aluminum.

No. 1970 —(3 doz. in box)... ... ... Each

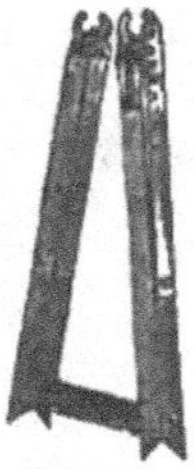

Closed.

Myers' Street Stand.

Street Stand, Myers'

This stand will hold eight bicycles in position and serve for a business sign at the same time. We furnish with plain sign, finished ready for lettering.

No. 1972........ .. Each

Truing Stand, Bicycle, Millennium

The screws have cupped centres, and will take any spindle. By far the best finished and most convenient stand on the market.

The truing stage is so arranged that the wheels can be trued from back or front of rim.

No. 1987—Weight, 26 lbs. .....Each

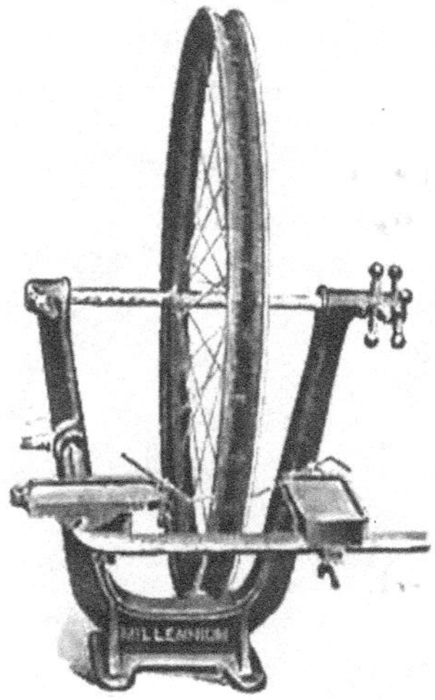

Truing Stand
Motorcycle, Millennium

This stand is suitable for bicycle or motorcycle wheels, and takes all sizes of spindles.

No. 1987—
Price, each

Soaps, Hand Cleaners

Gre-Solvent Uko.

More effective than soap in removing grease and grime. Much in demand by mechanics, laborers and munition workers.

Put up in 10 oz. attractive lithographed cans.

No. 1993 .... Per doz.

Soapstone

Put up in regular long, slim container, with sprinkler top.

No. 18070—1 lb sprinkling top (3 doz. in box) .... Each

No. 18152—1 lb. packages, bulk .... Per lb.

Use Plenty of Soapstone

A LIBERAL application of Soapstone sprinkled on your inner tubes will noticeably lengthen the life of the tires and give you better service from them.

The Soapstone reduces the friction which would otherwise take place between the tube and the canvas lining of the covers. It also acts as an insulator, protecting the inner tube from the heat generated in the tires when moving quickly under load.

Another trouble which is overcome by the use of plenty of soapstone is the tendency of the inner tube to stick to the cover lining, which "kills" that portion of the inner tube which comes in contact with the outer cover.

When making repairs, Soapstone should, of course, be sprinkled freely over any patches, as an extra precaution to prevent surplus cement from sticking to the cover.

Soapstone costs very little, and should be used freely.

Tubing, Steel

The following is a list of sizes on Tubing which have been used in the manufacture of C.C.M. Bicycles. Under ordinary circumstances we have a stock of these sizes at all times and can supply any quantity:

7/16" x 20 gauge	Per foot
1/2" x 20 gauge	Per foot
5/8" x 12 gauge	Per foot
5/8" x 20 gauge	Per foot
5/8" x 22 gauge	Per foot
3/4" x 18 gauge	Per foot
13/16" x 18 gauge	Per foot
7/8" x 18 gauge	Per foot
7/8" x 20 gauge	Per foot
1" x 16 gauge	Per foot
1" x 20 gauge	Per foot
1 1/16" x 20 gauge	Per foot
1 1/8" x 20 gauge	Per foot
1 3/8" x 20 gauge	Per foot

Other Sizes. In addition to the above, we have a limited quantity of other sizes of tubing, ranging from 11/16" to 1 5/8" in various gauges. These are not now used in the construction of C.C.M. Bicycles, but can be furnished at reasonable prices, subject to prior sale.

Torches

Hot Blast Torch—Gasoline

With attachments for holding soldering iron (not shown in illustration).

Capacity 1 qt., height over all 9½", diameter 4", weight 3 lbs., consumption ½ pt. per hour. Improved automatic pump.

No. 18045 ...Each

Taps, Crank Repair

Taper Taps, for tapping out cranks in standard size.

No. 504—½" x 20 thread (1 pr. in wooden box) ....Per pair

No. 528—$\frac{9}{16}$" x 20 thread (1 pr. in wooden box) ....Per pair

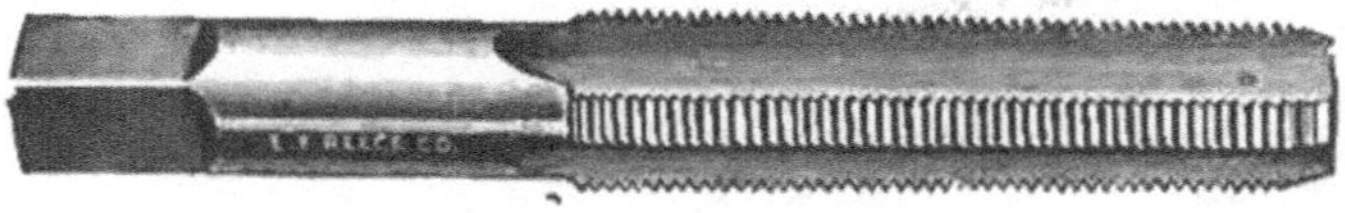

No. 504.

Tap and Die Sets
Bicycle

(Made in Canada)

A most convenient assortment. A complete outfit for ordinary repairs. The dies, being made in two parts, allow a large range of sizes to be threaded with one die.

All dies in this set are adjustable, and will cut $\frac{1}{16}$" larger or smaller than size marked on die.

No. 2100 (99)—Stock 12 inches; **14 dies**, cutting ¼", 24, 30, 32; $\frac{5}{16}$", 16 18, 20, 22, 24, 26, 28, 32; ½", 20, right hand; ½", 20, left hand. **23 Bicycle Taper Taps**, cutting ¼", 20, 22, 24, 26, 28, 30, 32; $\frac{5}{16}$", 18, 20, 22, 24, 26, 28, 30; ⅜", 16, 18, 20, 22, 24, 26, 28; ½", 20, right hand; ½", 20, left hand; including **one pair tap Wrench Dies.**

Price ....Each

Tap (Crank) and Reamer Set

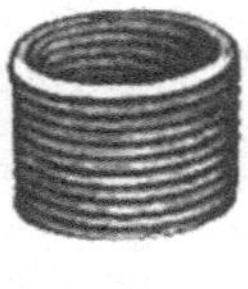

Bushing.

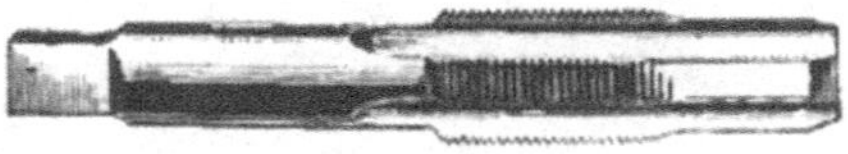

Tap and Reamer
Four in One.

Contains 1 pair ⅝ x 24 Taps, R. and L. Hand.

Six ½" x 20 R. H. Bushings.

Six ½" x 20 L. H. Bushings.

By drilling out the crank, tapping with ⅝ x 24 tap, and screwing the special bushings supplied on the pedal pins, a highly satisfactory and permanent job is made.

No. 507—Per set (1 set in box)..

No. 508R—Extra **Bushings, Right.** .... Each

No. 508L—Extra **Bushings, Left.** .... Each

Bushings packed 12 Rights and 12 Lefts in box.

MOTORCYCLE TAP AND DIE SET.

(Made in Canada)

Tap and Die Sets, Motorcycle

This set will thread all the bolts and nuts in the Yale, Indian, Harley-Davidson, Thor, Excelsior, and other makes of Motorcycles. The only combination put up strictly for this purpose. Consists of one 9" **Stock with six 1" dies**, cutting $\frac{5}{16}$", 24, 28; ⅜", 24; ½", 20, 24; $\frac{9}{16}$", 24. **One 5" Stock with six** ⅝" **dies**, cutting ¼", 20, 24; No. 6, 32; No. 8, 32; No. 10, 24, 32. **Ten taper hand taps**, cutting ¼", 20, 24, 26; $\frac{5}{16}$", 24, 28; ⅜", 24; ½", 24; $\frac{9}{16}$", 24; ½", 20 R. & L. **Four machine screw taps**, cutting No. 6, 32; No. 8, 32; No. 10, 24, 32, and one No. 9 tap wrench.

No. 2101 ....Per set

Useful Information Regarding Tires

We carry the guarantee—We agree with the purchaser of every guaranteed Dunlop or C.C.M. Lucky Tread Tire purchased from us to make good by repair or replacement when delivered to us **transportation charges prepaid,** any imperfection or defect in material or workmanship of such tire not caused by misuse or neglect.

For your convenience we have prepared a **Tire Adjustment Tag** which we will supply free on request.

This enables you to make your own adjustments on the spot, and we will make the same adjustment with you when goods are returned with tag attached.

We reserve the right to discontinue this privilege if it is abused.

The guarantee does not cover punctures, tubes damaged by being improperly fitted in their casings, or covers made unsatisfactory through being oil soaked.

Caution—All inner tubes should be partially inflated before being fitted. They are then less liable to become pinched.

We recommend the use of good grade tires, and while we sell the low-priced kind, too, we want to remind you that, **especially in rubber goods, quality counts.**

A generous application of **Soapstone** when fitting, will reduce inner tube troubles fifty per cent.

Canadian Standard Sizes

Measure straight across wheel from centre of underside. A wheel measuring:—

17 inches needs a 20-inch tire.
19 inches needs a 22-inch tire.
20½ inches needs a 24-inch tire.
22½ inches needs a 26-inch tire.
24½ inches needs a 28-inch tire.
26½ inches needs a 30-inch tire.

In Canadian Standard Size the circumference of the Dunlop rim is always the same—whether for 1¼, 1⅜, 1½, or 1⅝ tire.

Unless otherwise specially instructed, all tires are supplied for Canadian Standard Size Rims.

Instructions for Ordering Tires for English Bicycles

Unlike Canadian or American, on English bicycles, the circumference of the rim changes according to the size of tire fitted—the 1⅜ rim is greater in circumference than 1½, and the 1½ than the 1⅝. The circumference of English size steel rims for Dunlop wired-on tires is as follows:—

1⅝-78[illegible] inches; 1½-80[illegible] inches; 1⅜-81[illegible] inches.

This measurement is taken by passing a tape measure around the extreme edge of the rim—or by marking the rim—giving it one complete revolution on a board and taking the exact distance between the starting and finishing points.

In some cases the markings on the old tire will give this information.

When ordering, please specify whether covers are required for "Wired-on" or "Clincher" beaded edge rims.

Where not mentioned, we will supply the "Wired-on" Dunlop type.

First Class Freight Rate on Bicycle Tires

Bicycle tires are entitled to first class freight rate, but are frequently charged at 1½ times first class rate. This is because tires other than Bicycle take the higher rate in the classification. Bicycle tires, however, are entitled to a rating of first-class.

Please check over freight bills carefully and apply for refund in case of overcharge.

Dunlop Tires

(Made in Canada) **We Carry the Guarantee** See note on page 62.

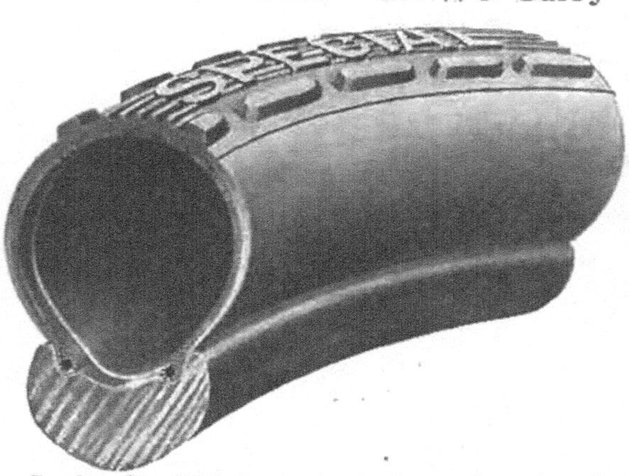

Dunlop Special.—Guaranteed 12 months. Standard equipment on C.C.M. Grade "A" Bicycles.
Complete, pair. Casings, each. Tubes, each.

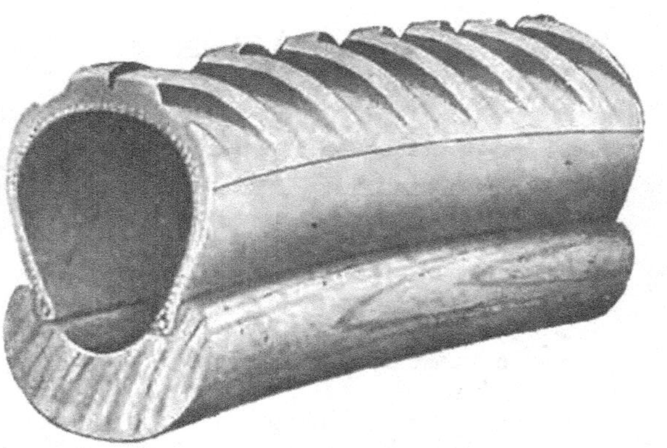

Dunlop Traction.—Guaranteed 12 months. Optional equipment on C.C.M. Grade "A" Bicycles. Standard equipment on C.C.M. Motorbike Bicycle.
Complete, pair. Casings, each. Tubes, each.

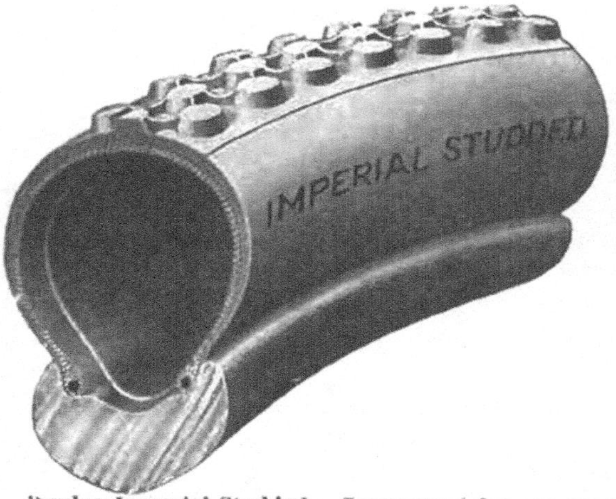

Dunlop Imperial Studded.—Guaranteed for one year. Standard equipment on C.C.M. Grade "B" Bicycles.
Complete, pair. Casings, each. Tubes, each.

Dunlop Peerless Studded.—Guaranteed 12 months.
Complete, pair. Casings, each. Tubes, each.

Dunlop Banner Studded.—Oval button-type tread Guaranteed for three months.
Complete, pair. Casings, each. Tubes, each.

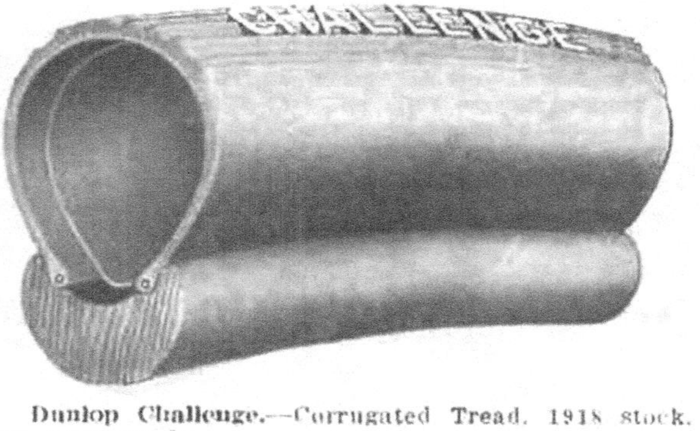

Dunlop Challenge.—Corrugated Tread. 1918 stock. Not guaranteed.
Casings, each.

NOTE.—We interpret the word "Tire" to mean Cover and Tube; if Cover only required, please specify.

"The Dealers' Interests Are Our Interests."

Dunlop Tire Price List

We carry the Guarantee (See note on Page 62)

Wire Edge Type Tires

		Complete Pair Tires	Casings Each	Tubes Each
Dunlop Peerless Studded, 28 x 1½" only	Guaranteed for 12 months			
Dunlop Traction, 28 x 1½" only	" " "			
Dunlop Special, 28 x 1½"	" " "			
Dunlop Special, English sizes, 28 x 1⅜" only	" " "			
Electric sizes, 28 x 1½"	" " "			
Dunlop Hercules Extra Heavy, 28 x 1⅝" and Extra Heavy Tube	" " "			
Imperial Studded, 28 x 1½", 26 x 1½"	Guaranteed for season			
Imperial Studded, 20 and 24 x 1½"	Guaranteed for season			
Banner Studded, Canadian size, 28 x 1½" only	Guaranteed for 3 months			
Banner Studded, English size, 28 x 1⅜" only	Guaranteed for 3 months			
Challenge, 28 x 1½" only. (No Challenge Tube made).	Not guaranteed			

Clincher Type Tires

		Complete Pair Tires	Casings Each	Tubes Each
Hercules Extra Heavy, 28 x 1⅝", with extra heavy Dunlop Special Tube	Guaranteed for 12 months			
G. & J., 28 x 1½" and 1⅝"	" " "			
Dunlop Special, 28 x 1½" only	" " "			
Imperial, 28 x 1½"	Guaranteed for season			
Banner (Old Style Tread), 28 x 1½" only	Guaranteed for 3 months			

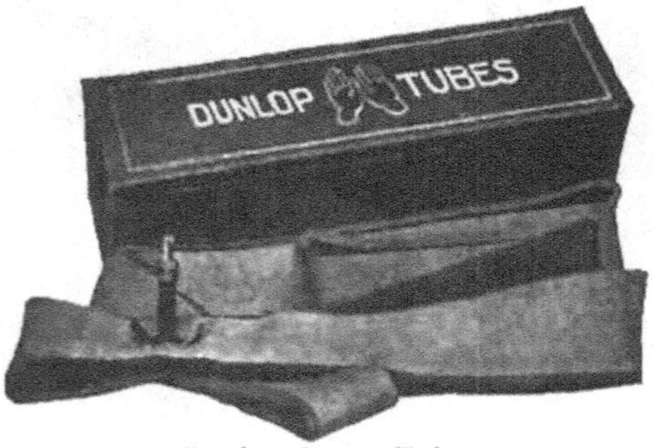

Dunlop Inner Tube.
Packed singly in cartons.

Tubes

(Made in Canada)

All Dunlop Tubes made from same material, but in different weights, which in 28" sizes are:

Dunlop Extra Heavy, 9 oz., guaranteed 12 months.
Dunlop Special, 8 oz., guaranteed 12 months.
Dunlop Imperial, 7 oz., guaranteed 1 season.
Dunlop Banner, 6 oz., guaranteed 3 months.

Dunlop Special Continuous, 28" (Guaranteed)	Each
Dunlop Extra Heavy Continuous, 28" x 1⅝" (Guaranteed)	Each
Clincher G. and J. Continuous, 28" x 1½", 1⅝" (Guaranteed)	Each
Dunlop Special Butt end, 28" (Guaranteed)	Each
Imperial Continuous or Butt end, 28" x 1½", 26" x 1½" (Guaranteed)	Each
Imperial Continuous or Butt end, 20" and 24". (Guaranteed)	Each
Banner Continuous, 28" x 1½" only (Guaranteed)	Each

NOTE.—Dunlop Special Tubes supplied with Peerless Studded, Traction Tread, Dunlop Special and Electric Tires.

Use plenty of Soapstone

Hints on Putting Tires on Rims

It is advisable to take the wheel out of the frame.

First see that the spokes do not project through the nipples and that the rim is properly taped.

Lay the wheel flat. Lay casing on rim. Inflate the tube slightly. Pull back cover at valve hole, place the valve in the hole and tuck tube inside the cover.

Draw valve out slightly, place cover in rim at valve hole, at the same time pushing the valve as far as possible through the rim.

Commence from the valve to place cover in rim, working half way round the rim to the right, then half way round the rim to the left, finishing at a point opposite the valve.

Inflate tube slightly and spin wheel to see that tire is on evenly all the way around before pumping the tire hard.

Prices Subject to Change Without Notice.

Tires (Continued)

"Lucky Tread"

Tires and Tubes

Guaranteed for the season. 28" x 1½" only.

Complete ....Per pair
Casings ....Each
Lucky Tread Tubes—28" x 1½" (guaranteed).
Continuous ....Each

Tires, Single Tube

Enduro

Enduro—Single Tube, 28" x 1½" ......Per pair

Brigadier

Brigadier—Single Tube, 28" x 1½" ....Per pair

"Record" Studded

Guaranteed 60 days.

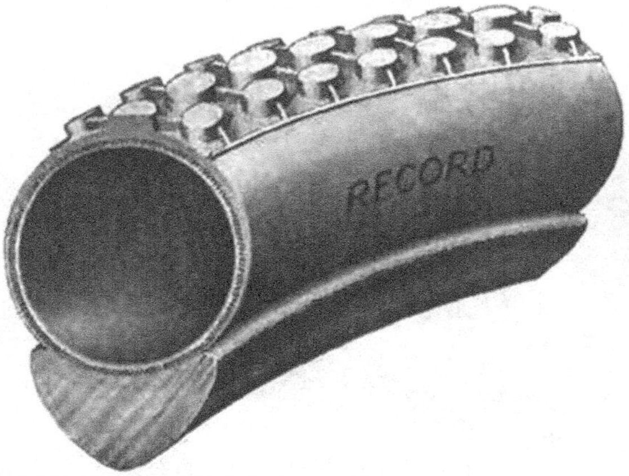

Record—Single Tube, 24" x 1½" ......Per pair
Record—Single Tube, 26-28" x 1½" ....Per pair
Record—Single Tube, 30" x 1½" ......Per pair

We do not stock Record Double Tube Tires.

"Reliance" Studded

(Not Guaranteed.)

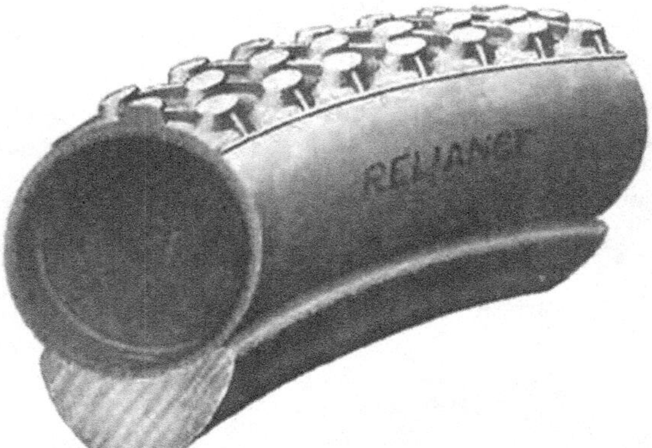

Size 28" x 1½" ....Per pair

Palmer

(Not Guaranteed.)

Palmer Single Tube Tires are supplied in two styles—Road and Light Racing or Track.

Track Racing—26" x 1" x 1⅛" x 1¼". Per pair
28" x 1" x 1⅛" x 1¼". Per pair
Road Racing—26" x 1⅛" x 1¼" x 1⅜". Per pair
28" x 1⅛" x 1¼" x 1⅜". Per pair

Tires (Continued)

"Juvenile"

Goodrich Single Tube—These tires have given excellent satisfaction. Supplied as follows:

	Per pair.
20" x 1⅜"	.....
24" x 1⅜"	.....
26" x 1⅜"	.....

Tires, Carriage, Dunlop Pneumatic

Clincher and Wire Edge Type. (Not Guaranteed.)

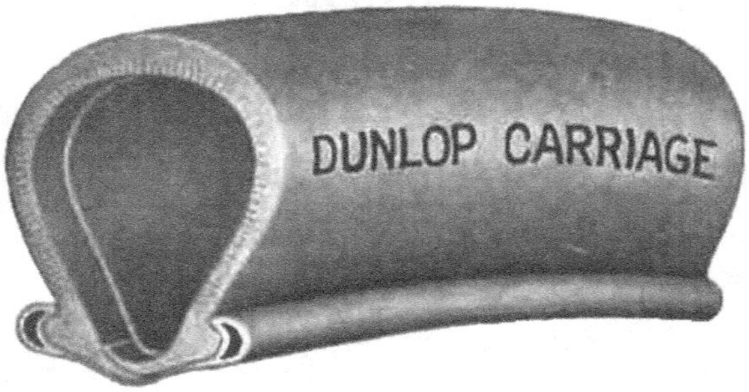

	Tires Complete Each.	Covers Each.	Tubes Each
Runabout—36" x 1¾" ...			
34" x 1¾" ...			
Stanhope—36" x 2"			
34" x 2"			

Butt End Carriage Tubes same price as Runabout and Stanhope in their respective sizes, viz.: 1¾", 2".

29¾" Rim requires 32" Tire. 31¾" Rim requires 34" Tire. 33¾" Rim requires 36" Tire.

When ordering please specify whether Tire is required for Wire Edge or Clincher Type Rim.

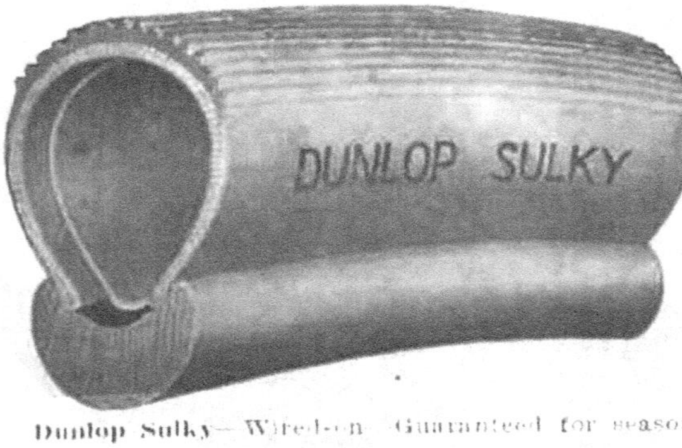

Dunlop Sulky—Wired-on. Guaranteed for season—

	Tires Complete Per Pair.	Covers Each.	Tubes Each.
26" and 28"			

Tires

Dunlop Motorcycle

Traction Tread

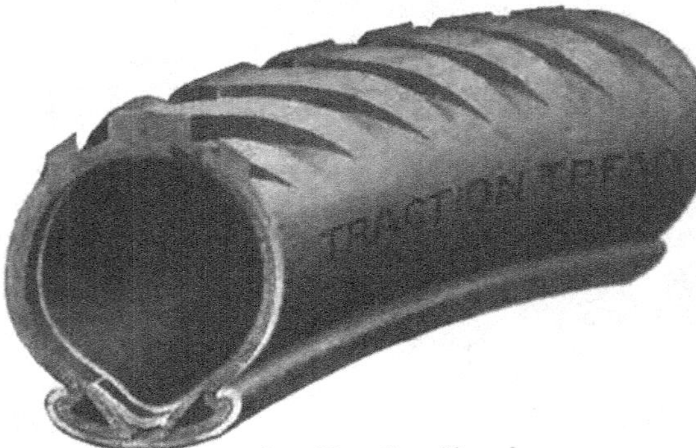

Dunlop Traction Tread.

Clincher Type, to fit all standard American Rims.

Size
Inches.
*26 x 2¼ for 1912 Rims.
*26 x 2½ for 1912 Rims.
28 x 2¼ for 1912 Rims.
28 x 2½ for 1912 and B. B.—O. Rim.
28 x 2¾ for B. B. and C. C.—M. Rim.
28 x 3 for C. C. M. Rim. Extra Heavy Service Type.
*Also made to fit English size rims.

Dunlop Rubber Studded.

Clincher Type, for 1912 American Rims only.

In following sizes:—
26 x 2¼, 26 x 2½, 28 x 2¼.

Studded

Wire Edge Type, Rubber Studded Tread.

For Canadian Rims only. For Light Machines. Regular Equipment on C.C.M. Motorcycles. **Size 28 x 1¾.**

NOTE.—**When ordering 28" tires state whether required for B. B. or C. C. Rims.** B. B. Rims are 1.50 inches (1½") wide between clinches, and C. C. Rims are 1.65 inches (1 13/20") wide between clinches.

Owing to the unsettled conditions of the rubber market, and to the fact that our catalogue is in your hands for a year, we are not giving you prices on motorcycle tires. All prices are subject to change without notice, and we sell at all times at the current Dunlop quotations.

Tire Deflator

Saves time and trouble, as you will avoid getting dirt into valve.
No. 2200 .... Each

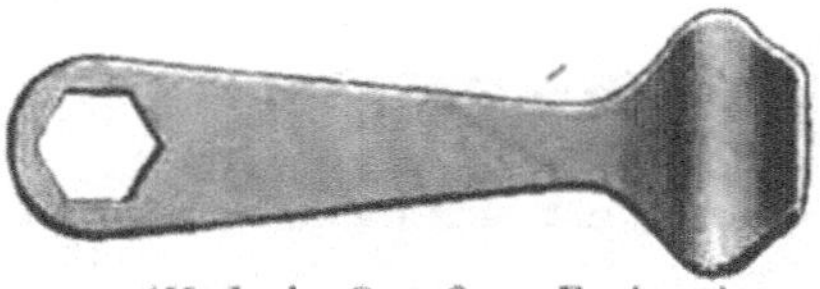

Tire Levers

(Made in Our Own Factory)

Tom Thumb—Will remove the tightest cover from rim with ease. Will not damage rim or injure tire. Nicely finished. For best results use two levers at once.
No. 2161—Per doz. ....

Tire Tape

C. C. M. *(Made in Canada)*

Made in the best canvas backing with pure Para rubber, and warranted to retain its adhesive qualities for one year. Supplied in 1 oz., 2 oz., and ½ lb. rolls.

No. 2130—C. C. M., 1-oz. rolls, 16 in 1-lb. carton .... Per lb.
No. 2131—C. C. M., 2-oz. rolls, 8 in 1-lb. carton .... Per lb.
No. 2132—C. C. M., ½-lb. rolls .... Each

Dunlop Electric Reliance

(Made in Canada)

No. 2150— 1-oz roll, 16 in 1-lb. carton ... Per lb
No. 2151— 2-oz roll, 8 in 1-lb. carton ... Per lb.
No. 2152— ½-lb. roll .... Each

Tape, White Rim

For taping inside of rim, covering spoke ends, to protect tube. Put up in rolls of 72 yards.

No. 2133... Each

(Some of our stock is in 36-yd. rolls, two of which will be supplied in place of one 72-yd. roll.)

Tire Repair Bands

(Bicycle)

Have two flaps, which fit over the wire edge and under the rim, keeping the band from creeping.

For repairing cuts, blow-outs, etc., on outer cover. Makes a smooth and permanent repair. Width 2" and 4". Two dozen in bundle.

No. 1650—2" x 4" .... Per dozen
No. 1651—4" x 4" .... Per dozen

Tire Sleeve, Motorcycle

Tough, flexible; wear like iron; fitted with rawhide strap for attaching. A necessity for blowouts, cut or weakened tire. **Supplied in plain tread only.**
No. 2180—6" x 7", with straps .... Each

Tire Band, Non-Skid Motorcycle

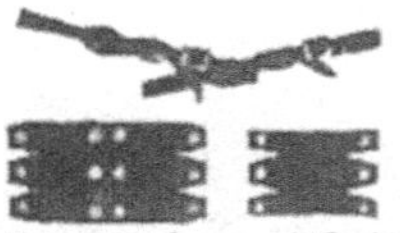

Set of six, for placing at regular intervals on rear wheel; will prevent skidding or slipping in mud, snow, slush, etc.

Directions—Strap the sleeve, or non-skid tire band, in desired position and as tightly as possible before tire is fully inflated. Complete inflation and tire will be ready for use.
No. 2190—Size 3½" x 7" each, with straps .... Per set of six

Tire Levers, Motorcycle

One special shaped lever starts tire away from rim, each of other two digs under bead of tire, pulls bead over rim and hooks on to spoke. Levers 7" long, dull finish. Packed in enamel duck case.
No. 2162 (710)—Per set of three ....

No. 710 Set.

Tiring, Baby Carriage

North Pole

This brand of Baby Carriage Tiring is very popular with repair men, because it can be easily fitted in much less time than the regular kind, and requires no tire fitting machine. There is positively no waste, and it will not gape open at joint. Will not creep on rim, and having no hollow centre, wears longer. In addition to its unique mechanical features, it is made of high quality rubber, which ensures long life.

No. 2235—⅜", with Joining Pins, 60 feet in roll. Weight per roll 6¾ lbs. ....Price, per foot

No. 2236—$\frac{7}{16}$", with Joining Pins, 60 feet in roll. Weight per roll 7 lbs. 10 oz.....Price, per foot

No. 2237—½", with Joining Pins, 60 feet in roll. Weight per roll 9 lbs. 3 oz.....Price, per foot

No. 2238—⅝", with Joining Pins, 60 feet in roll. Weight per roll 15 lbs.........Price, per foot

No. 2239—¾", with Joining Pins, 60 feet in roll. Weight per roll 22½ lbs.........Price, per foot

How to Fit

1. Place tire on bed of rim to get length. In case of a 25" wheel, cut rubber, with a sharp knife, 3" short of the actual circumference. Sever the steel centre with a cold chisel, by placing the cut portion over a piece of steel.

2. Place one of the steel joining pins supplied, 1¾" long, in a vise, leaving half of it projecting. (Pins 1¾" long are supplied for ⅜", ½", $\frac{7}{16}$" and ⅝" tire, whilst ¾" and larger sections take extra thick pins, 2¼" long.)

3. Take one end of the tire already cut, and push same on to the projecting joining pin until the half (⅞") is inserted.

4. Reverse, placing rubber firmly in vise so that the remaining half of joining pin projects from vise, and then repeat method described in paragraph No. 3 with other end of tire.

5. The tire is now complete and may be sprung on to the wheel, taking care to place the joint on the rim first, and stretching evenly all round.

It is important that tires should not be rolled on to rim and left in twisted (unnatural) condition. Strict observance of this will result in greatly increased wear.

Gripping Pliers and Joining Pins for North Pole Tiring

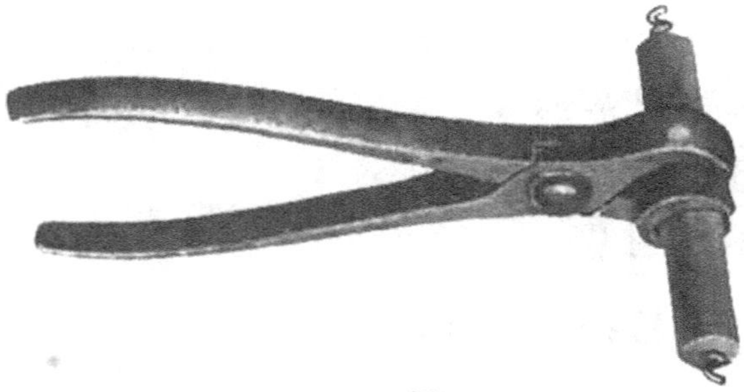

Gripping Pliers.

No. 2245—Gripping Pliers for use in fitting North Pole Tiring ....Each

Instructions When Using Gripping Pliers on North Pole Baby Cab Tire.

"In order to obtain a secure grip, it is always advisable to insert a piece of thick cloth (a piece of old cycle cover answers admirably) between the rubber and pliers, as this not only increases the surface to be gripped and so gives a securer hold, but also protects the tiring from damage from the pliers."

Joining Pins

No. 2240—Extra Joining Pins for sizes ⅜", $\frac{7}{16}$", ½" and ⅝" ....Per 100

No. 2241—Extra Joining Pins for size ¾" ....Per 100

Tiring, Baby Carriage, Regular

Manufactured from superior quality rubber, very resilient, making a smooth, easy running wheel. Will stand a lot of hard service.

This Tire is exceptionally good value at a moderate price, and is very serviceable. Unguaranteed.

	No. ft. to pound.	Per foot.
No. 2220—⅜ x 7/64 hole. 60 ft. in roll...	10.81	
No. 2221—$\frac{7}{16}$ x 7/64 hole 60 ft. in roll...	8.16	
No. 2222—½ x 7/64 hole. 60 ft. in roll...	5.79	
No. 2223—⅝ x 7/64 hole. 60 ft. in roll...	3.5	
No. 2224—¾ x 1/8 hole. 60 ft. in roll...	2.56	
All sizes ....Per lb.		

Tire Wire, Baby Carriage

No. 2230—No. 16 gauge, half-hard, galvanized ....Per lb.

Valves

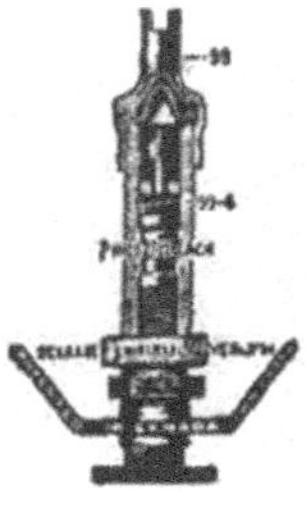

Dunlop Metal Base—This valve is used on all Dunlop Tires. It is positively air tight.

No. 2501..Per 100

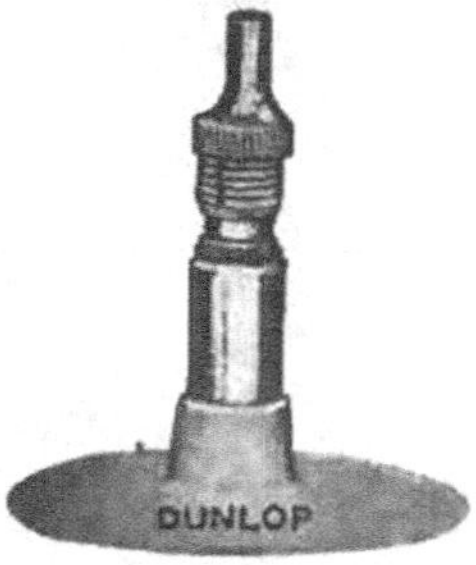

Schrader Rubber Base Complete with stem.

No. 2503...Per doz

Valve Parts, Schrader

Valve Cap Rubbers.

No. 2520....Per 100

Valve Caps.

No. 2521—(500 in box) ....Per 100

No. 21101 — Heavy Valve Cap (Auto) ....Per 100

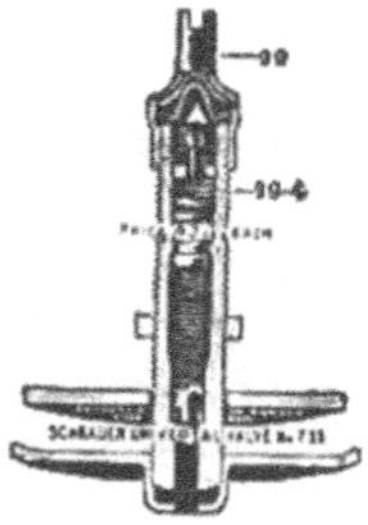

Oval Base.

This valve has a detachable base, and makes a much better job than the regular metal base, as it has a larger surface on the tire. Owing to its shape, it cannot pull loose or cut, even if the tire creeps, and will hold firmly in tires full of repair composition.

No. 2505 — (Small) ....Per 100

No. 2506 — (Large) ....Per 100

Valve, Dunlop Motorcycle

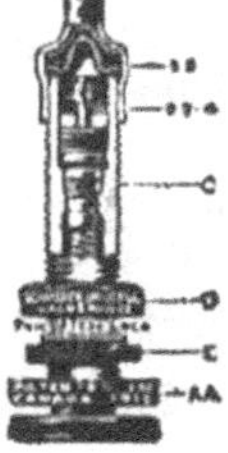

The valve shown here is the well-known Schrader line, universally recognized as standard. All Dunlop inner tubes are equipped with it.

No. 2500 ..Per 100

Valve Insides

Schrader Universal Valve Interior.

Used on all types of pneumatic tire valves.

No. 2535—(5 in tin box, and 20 tin boxes in a carton)...Per 100

Valve Tool---C. C. M.

C.C.M. Valve Tool (4 in 1)—A very useful article, which will execute four repair operations on a valve.

No. 19101—(25 in box) ....Each

Varnish

Air Drying and Baking.

Supplied in one-pint tins. Imperial measure.

No. 2553—Air Drying ....Each

No. 2541 — Baking .... Each

Vaseline

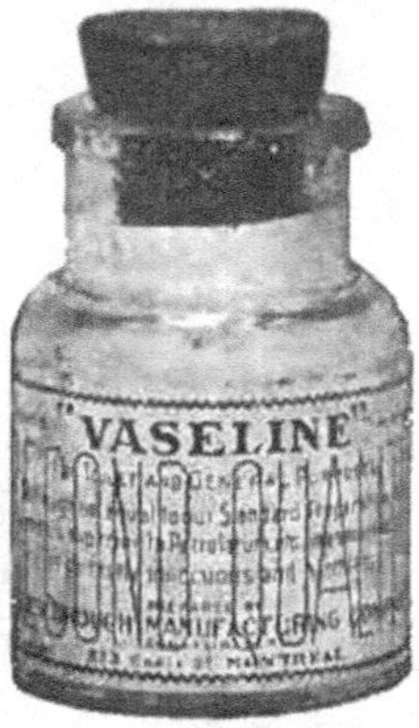

Pure Jelly Petrolatum is attractively put up in bottles and 1-lb. cans.

No. 2550 — 5c. bottles (1 doz. in box) ...Per dozen

5c. bottles..Per gross

No. 2551—1-lb. cans .... Each

Washers
"Key" and "D," Nickelled

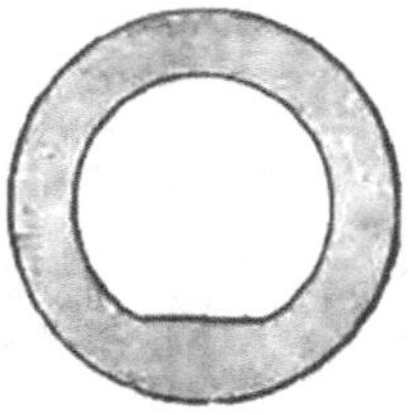

"D" Washer.

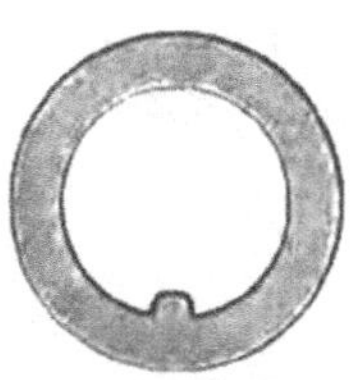

"Key" Washer.

These washers are used on the crank axle. They are furnished in the following sizes:—

⅝", 11/16", ¾", 13/16", ⅞", 1".

Cannot obtain ¾" "D" Nickelled Washers.

⅝", 11/16", ¾" ....Per 100

13/16", ⅞", 1" ....Per 100

Washers (Continued)

Round Nickelled Steel Washers.

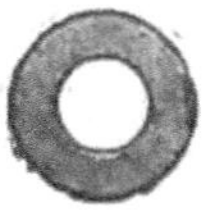

Nickel-plated. They can be furnished in the following sizes:—

$\frac{3}{16}$", $\frac{1}{4}$", $\frac{5}{16}$", $\frac{3}{8}$", $\frac{7}{16}$", and $\frac{1}{2}$".

100 in package ....

Washers, Spoke Nipple

Steel Rim

No. 2650—(1,000 in box) ....Per 1,000 $2.40

Washers, Wood Rim Spoke Nipple

Plain.

No. 2651—Saw Tooth (1,000 in box)...Per 1,000

No. 2652—Plain (1,000 in box) ......Per 1,000

Washers, Lock

Put up in a strong tin box of assorted sizes, from $\frac{1}{4}$" bolt size to $\frac{5}{8}$" bolt size, advancing by sixteenths of an inch.

No. 22000—Price per box .....

The following sizes sold only in boxes of one hundred:—

$\frac{3}{16}$ $\frac{1}{4}$ $\frac{5}{16}$ $\frac{3}{8}$ $\frac{7}{16}$

Sold only in boxes of fifty.—

$\frac{1}{2}$ $\frac{5}{8}$ $\frac{3}{4}$

Wheels, Built-up (Less Tires)

We are offering to the trade, wheels complete, less tires. Fitted with Hercules spokes, Dunlop or Crescent rims and hubs as below.

All built-up wheels supplied with 28" Canadian size rims unless otherwise specified.

We also carry **sulky wheels** complete, less tires, fitted with knock-out axle hubs, Dunlop rims.

No. 2700—Front fitted with **No. 1047 (25)** Hub. .... Each

No. 2701—Front fitted with **No. 1036** Comfort Hub .... Each

No. 2750—Rear fitted with **No. 1030** Comfort Hub (state sprocket required) ....Each

No. 2751—Rear fitted with **Hercules Coaster Brake No. 308.** (State sprocket required).Each

No. 2800—Wheel fitted with **Sulky Knock-Out Axle Hub, No. 1040** ....Each

Rear Wheels packed 9 to crate.

Front wheels packed 12 to crate.

Front and rear wheel when ordered with striped rims, as listed on page 52, add difference between natural finish and striped rim.

NOTE.—Wheels ordered with striped rims have to be built up after order is received, which causes some delay.

In ordering, specify whether required with Dunlop or Crescent rims.

Owing to extreme shortage of steel rims we have discontinued cataloguing **Built-up Wheels with Steel Rims.**

Best Method of Building-up Bicycle Wheels

TO AVOID confusion it is well to lace the nine inside spokes on one side of the hub first. Drop these spokes into the outside countersunk holes of one end of hub. Hold with spoke side of hub toward the operator. Insert spoke into second nipple-hole to right of valve hole. Leave the next three holes to right. In the fourth hole insert the next spoke. Continue in the same way until the nine spokes are laced.

Then turn the wheel with other end of hub upwards and drop the nine outside spokes into their holes. Drop the nine inside spokes into outside countersunk holes on this side. Each spoke will go into the nipple hole to the right of the opposite spoke already laced, commencing opposite the first spoke laced and working to the right until the nine are done.

This will leave the nine outside spokes hanging loose on the side on which we started. Lace these, each one crossing three spokes to the right and being inserted into the second vacant nipple hole to the right.

Into the nine holes left, drop the nine outside spokes. Turn the wheel over and lace these, each one crossing over three spokes and fitting into the nipple holes left vacant for them.

It will be noted that the holes on one end of hub are not drilled exactly opposite to the holes on the other end, but opposite a point midway between them. This is to allow for the spoke reaching to the rim nipple-hole to the right of the hole taking the spoke from the opposite end of the hub and avoids the necessity of using more than one length of spoke. This applies to both inside and outside spokes.

In tightening the spokes run down each nipple until lower end leaves only last thread of the spoke exposed. To complete tightening be careful to give each nipple the same number of complete turns, in order to keep the wheel true.

Wrenches

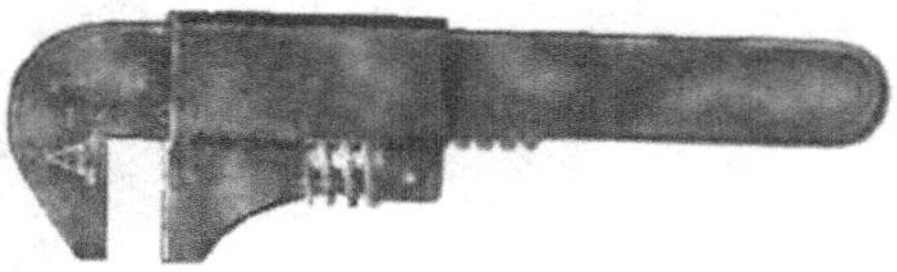

No. 2905-2908.

No. 2905—A well-made wrench of neat design, solid jaws. Opens 1¼ inches. **Nickelled and polished.** (12 in box) ....Per dozen

No. 2908—Same as **No. 2905**, but in **unpolished nickel** finish. (12 in box) ....Per dozen

No. 2901 (103)—A well made wrench with drop forged solid jaws, **nickel-plated**. Length 5½", jaws ¾" x ¼", opens 1½". Weight 5½ oz. ....Per doz

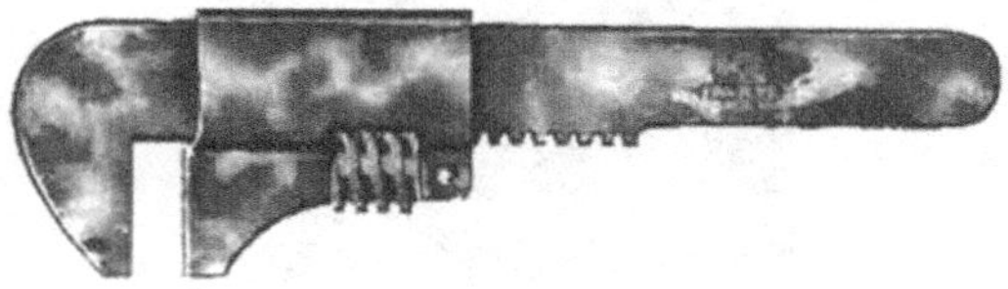

A 2.

Length 5½". Opens 1⅝". **Mottled** finish.

No. 2906 (A2)—(12 in box) ....Per dozen

A3—Tire Lever Wrench.

Length 6". Opens 1¼". **Nickel-plated.**

No. 2902 (A3)—(12 in box) ....Per dozen

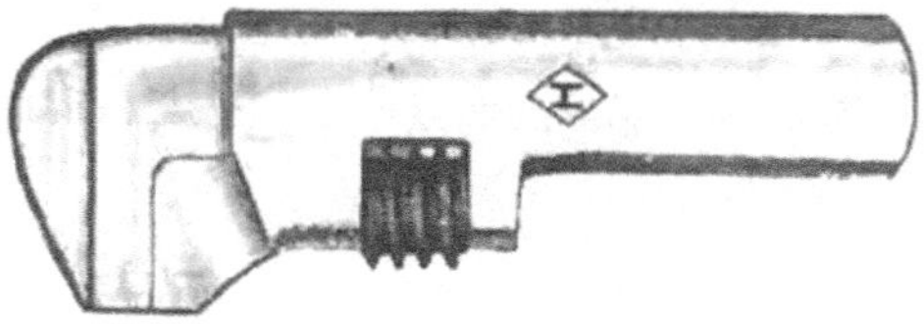

No. 22012 (105)—A handsome, high grade wrench, all steel construction, drop forged, **mottled** finish. Length 4¼", jaws ¾ x ¼", opens 1½". Weight 6¼ oz. ....Per dozen

Wrench, Pedal, Shop

Specially adapted for attaching and detaching pedals. Opening ⅝".

No. 2907 ....Dozen

Wrench, B. & S. Shop

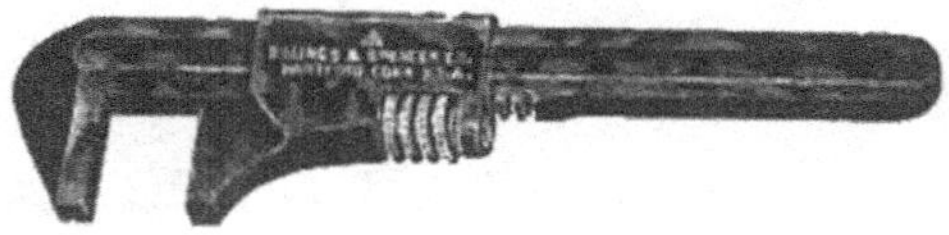

B. & S.

B. & S. Shop Wrench—Total length 8". Opens 1½". Thickness of jaw ⅜" Depth of jaw 1¼". Weight 16½ oz.

For use on bicycles, motorcycles, automobiles, etc. Drop forged from bar steel. **Semi-finished.**

No. 2912 ....Each

Wrench, Sprocket

No. 2913.

Sprocket—To remove rear sprockets from the hub. Made from malleable iron and fitted with six inches of bicycle chain. $\frac{3}{16}$ x ½ pitch (suitable for ½" or 1" pitch sprockets). Enamel finish.

No. 2913 ....Each

Wrenches (Continued)

Motorcycle Wrench

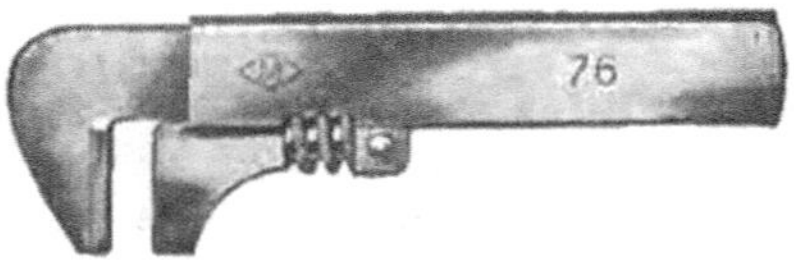

Patent Telescope Wrench.

Finest steel construction. **Mottled finish.** Dimensions: When closed 6", open 8"; jaws $\frac{3}{16}$ thick, $1\frac{1}{16}$" deep. Weight $10\frac{3}{4}$ ounces.

No. 2914 (76)—Each in carton (12 in box) ....Per dozen

Socket Wrench Set

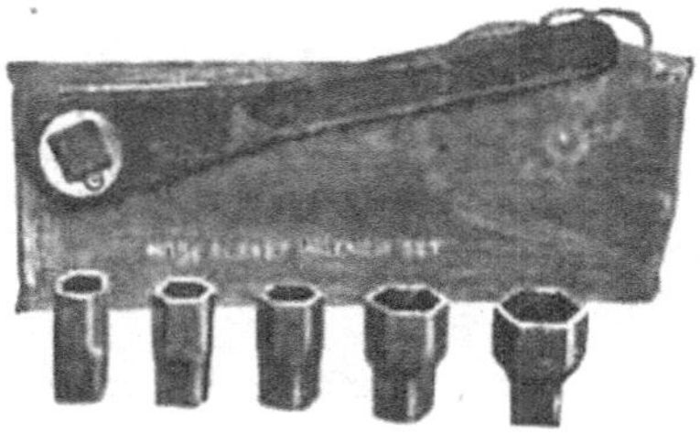

Consists of a ratchet handle and five sockets for hexagon nuts, $\frac{15}{32}$, $\frac{17}{32}$, $\frac{19}{32}$, $\frac{21}{32}$, $\frac{23}{32}$. Semi-finished. Put up in a strong enamelled duck case.

No. 2915 ....Per set

Engineer's Wrench Set

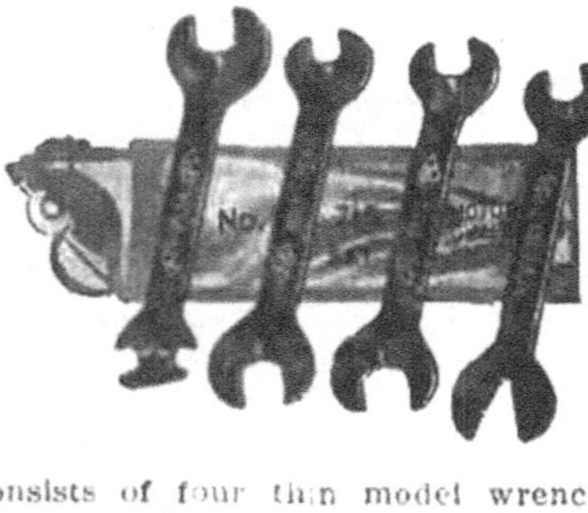

Set consists of four thin model wrenches, having nine different openings, as follows:—$\frac{5}{16}$, $\frac{3}{8}$, $\frac{15}{32}$, $\frac{1}{2}$, $\frac{9}{16}$, $\frac{5}{8}$, $\frac{11}{16}$, $\frac{3}{4}$ and Alligator. Packed in canvas kit.

No. 2916 (715) ....Per set

Tool Kit

We have selected for this kit an assortment of tools that are light but strong, and nicely finished. The hammer is drop forged in one piece, and the end of the handle is shaped into a tire tool.

No. 2250.

The Kit consists of the following tools in a leatherette case with straps.

- 7" Black Monkey Wrench.
- 5" Nickel-plated, Adjustable Combination Pliers.
- 2" Square Shank Screw-driver.
- 6" Flat File, with handle
- 4 oz. Hammer and Tire Tool.
- $\frac{3}{8}$" Drift Punch, Half Polished.
- $\frac{3}{8}$" Cold Chisel, Half Polished.
- 1 Box Assorted Cotter Pins.

No. 2250 ....Each

Bicycle Tool Set

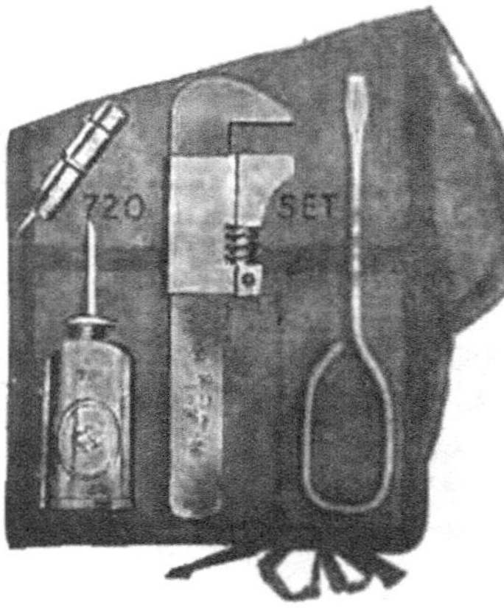

Contains one wrench (No. 2908), unpolished, nickel finished.

One oiler (No. 1376), nickel-plated.

One wire screwdriver (No. 655), carefully tempered and nickel-plated.

Wrapped in neat brown canvas tool-roll.

No. 2918 ....Per set

Whistles

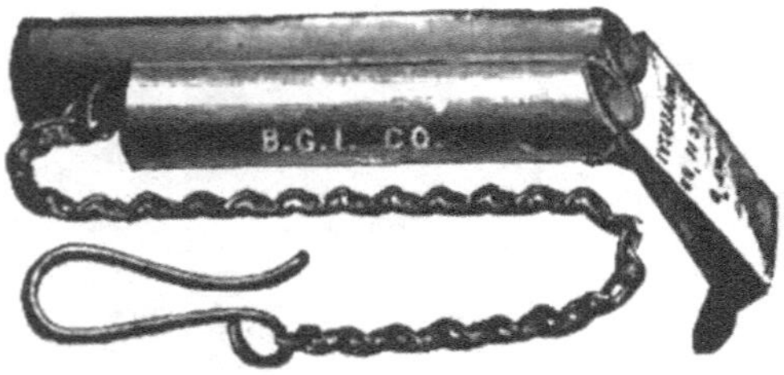

Universal

No. 2670—It is nicely nickel-plated (1 dozen in box) ......

Per doz., **$5.20**

The Echo Call

No. 2671—Made of brass, highly finished and nickel - plated (1 doz. in box)

Per doz

Prices Subject to Change Without Notice.

AUTOMOBILE ACCESSORIES

Columbia Batteries

Columbia Multiple, Multiplex and Hot Shot Batteries

There are a great many places where a unit battery is desirable, and often conditions are found which preclude the use of a single series set of coils, or even a single series set of coils loosely connected. In this case a series multiple battery of dry cells, or a unit battery of even a single set of cells in series will successfully accomplish the work.

Such service is represented by boat lighting, electric vibrator massage machines, refrigerator lights, auto horns, etc.

(Made in Canada)

Type 356.

Columbia Multiple Battery.

Every cell is insulated from the case and from the other cells in such a way as to make a waterproof and practically damage-proof unit battery. Enclosed in a hermetically sealed, handsomely enamelled **metal case,** with convenient handles for carrying.

(Not carried in stock at Winnipeg Branch.)

Type	Voltage	Length in.	Width in.
356	7½	13¾	8⅛

Height over all in.	Weight lb.	Price
8¾	50	

Type 356.

Columbia Multiplex Battery.

The only difference between this battery and Columbia Multiple is that the Columbia Multiplex is furnished with a **wooden case.**

(Not carried in stock at Winnipeg Branch.)

Type	Voltage	Length in.	Width in.
356	7½	14¼	8⅝

Height over all in.	Weight lb.	Price
8	44	

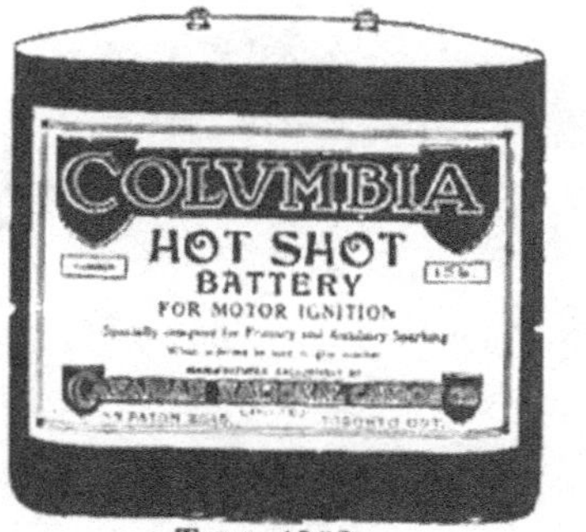

Type 1562.

Columbia Hot Shot Battery.

A unit battery in an attractive red, moisture-proof case. Particularly useful for ignition, where the variety of sizes makes it adaptable for motors of all sorts—stationary, vehicle or motor boat. It is very satisfactory for carriage lighting outfits.

Type	Voltage	Length in.	Width in.
1561	7½	13¼	2¾
1562	7½	8	5

Height in.	Wt. of Std. Pkg. lb.	Std. Pkg.	Price
7½	284	20	
7½	274	20	

Explanation of Tables

The type number indicates the number of cells in the battery and the voltage; and if there is a fourth figure, it indicates whether the cells are arranged in one row or two.

A 1562 Battery is, therefore, made up of one cell in multiple, giving a short circuit current of about 25 amperes; 5 cells in series, giving a voltage of 7½. The battery is made up of 6-inch cells, and is made in two rows.

Columbia 3-Volt Battery

(Made in Canada)

With Connections on Top.

Can be used for any three volt lighting on cars and to fit any commercial hand lantern now on the market. We are prepared to furnish a three-volt battery in the same size as our regular No. 6 cell. Care should be taken to use a three-volt bulb with this battery. When ordering state whether required with or without connections on top.

(Not carried in stock at Winnipeg Branch.)

No. 1015—Price, in lots of less than $5.00 worth at wholesale prices ....Each

No. 1015—Price, in lots of from $5.00 to $75.00 worth at wholesale prices ....Each

No. 1015—Price, in lots of over $75.00 worth at wholesale prices. .... Each

Without Connections on Top.

Columbia & Columbia Ignitor Dry Cells

(Made in Canada)

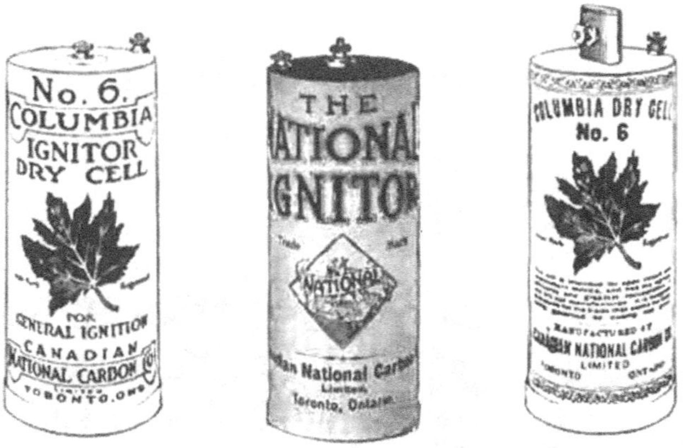

Columbia and Columbia Ignitor Dry Cells Furnished in Square Cartons or with Fahenstock Connections if Desired.

		Sizes of Zinc Cans	Price each Per Bbl.	Price each Broken Lots
No. 1010—National Ignitor	No. 6	2½" x 6"		
No. 1013—Columbia Ignitor	No. 6	"		
No. 1014—Columbia Regular	No. 6	"		

Future Delivery Prices Not Guaranteed

We cannot guarantee prices for future delivery on Columbia or National Batteries; but we will accept Battery orders for future delivery at prevailing prices at time of shipment.

Carbide

C.C.M.

Put up specially for automobiles and motor boats. The best brand Calcium Carbide.

No. 2030—5 lb. tins. (1 doz. in wooden case.) Price, doz. ....

No. 2028 — 100 lb. drums. Price, each ....

Cement

M. & M.

A patch applied with ordinary cement to an inner tube is not a permanent repair, but a patch applied with M. & M. Cement will effect a perfect union. Exceptional quality.

No.	Size	Price Per doz.
2330—	¼ pint	......
2331—	½ pint	......
2332—1	pint	......
2333—1	quart	....

Chain Lever Hooks

This special **Rim Chain Connector** is used on both "C.C.M. Dreadnaught" and "Maxim Imperial" Chains. This connector is the only type which allows the user to adjust chain properly with a minimum of effort. Positively prevents loss of chains by becoming disconnected.

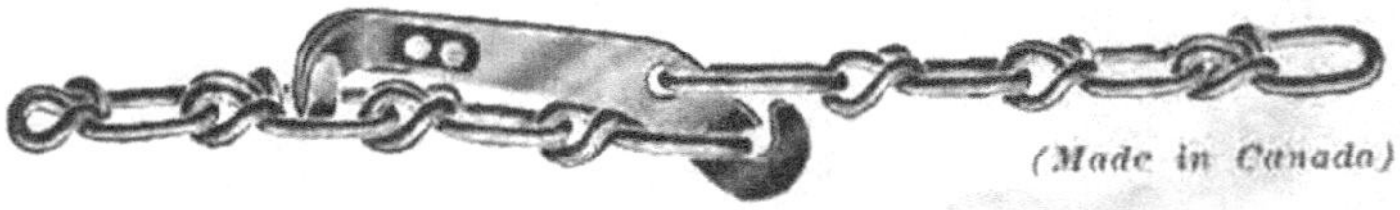

(Made in Canada)

No. 2813—Right, per 100 ....

No. 2814—Left, per 100 ....

Prices Subject to Change Without Notice.

Chains--"C.C.M. Dreadnaught" Tire Chains

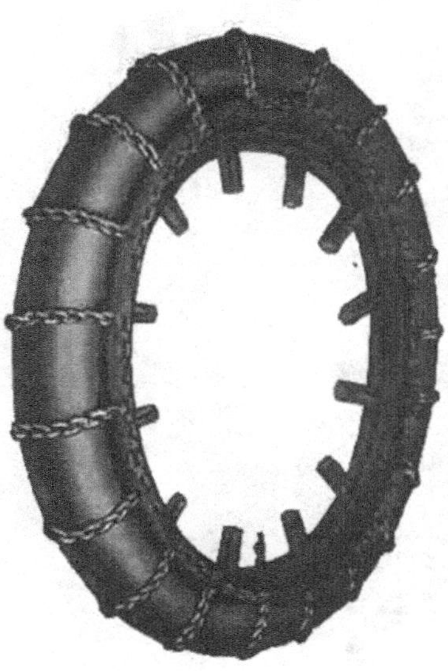

(Made in Canada)

Although lower in price than some others on the market, they are made to a quality standard that cannot be excelled, regardless of price.

The Electric Welded Cross Sections, after being inspected and tested, are carefully bone-hardened, resulting in links with a diamond-hard surface and tough inner core, thus assuring a maximum of strength and wear.

The Rim Chain is galvanized and Cross Chains copper plated to prevent rust.

A special Lever Rim Chain connector, on which is stamped "Dreadnaught," is used on these chains. (See illustration below.)

Dreadnaught Chains are guaranteed as to quality of material and workmanship, and to give at least same service as the highest priced chain on the market.

Packed one pair in a strong canvas bag.

No.	Size	Code Word	Per pair	No.	Size	Code Word	Per pair
2074	30 x 3½	Slip		**2083**	35 x 4	——	
2079	31 x 4	Positive		**2089**	34 x 4½	Not	
2070	32 x 3	On		**2090**	35 x 4½	——	
2075	32 x 3½	Hills		**2084**	36 x 4	We	
2080	32 x 4	Or		**2091**	36 x 4½	Can	
2081	33 x 4	Mud		**2098**	36 x 5	Deliver	
2076	34 x 3½	You		**2092**	37 x 4½	——	
2082	34 x 4	Are		**2099**	37 x 5	At	

"Maxim Imperial" Tire Chains

While better value is obtained by purchasing C.C.M. Dreadnaught Chains, Maxim Imperial Chains are satisfactory **for light cars**, and purchaser will obtain full value.

The special Lever Rim Chain Connector is supplied. (See Illustration below.)

The Rim Chain is **1/32" heavier** than supplied on other chains of this grade, giving the chain greater durability.

The Cross Sections are polished and not hardened. **None of the low priced chains have hardened Cross Sections.**

Serial No.	Size.	Price per pair	Serial No.	Size.	Price per pair
2522	30 x 3½		**2532**	36 x 4	
2523	32 x 3½		**2537**	34 x 4½	
2524	34 x 3½		**2538**	35 x 4½	
2527	31 x 4		**2539**	36 x 4½	
2528	32 x 4		**2540**	37 x 4½	
2529	33 x 4		**2546**		
2530	34 x 4		**2547**		
2531	35 x 4				

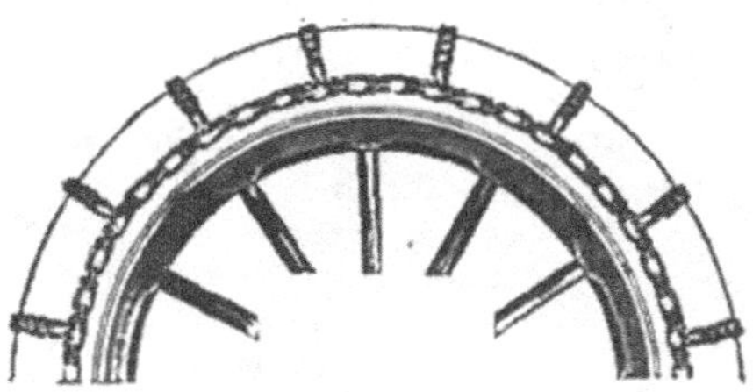

(Made in Canada)

Maxim Imperial Repair Cross Chains

(Made in Canada)

These Cross Chains are copper plated to prevent rust.

Per 100

No. 2788—3½" .....
No. 2789—4" ....
No. 2790—4½" .....
No. 2791—5" ....

C.C.M. Dreadnaught Repair Cross Chains

(Made in Canada)

Electrically welded and carefully bone-hardened, resulting in links with a diamond-hard surface and tough inner core, assuring maximum strength and wear. Copper plated to prevent rust.

Prices per 100

No. 2151—3" ....
No. 2152—3½" ....
No. 2153—4" ....
No. 2154—4½" ....
No. 2155—5" ....

Chain Repair Links

(Made in Canada)

For repairing broken Cross Sections.

No. 11082—Per 100 ...

Maxim Spark Plugs

(Made in Fuse Department of Russell Motor Car Company's Plant.)

No. 1490—½ inch (Three - in - One Terminal. Price, each .....

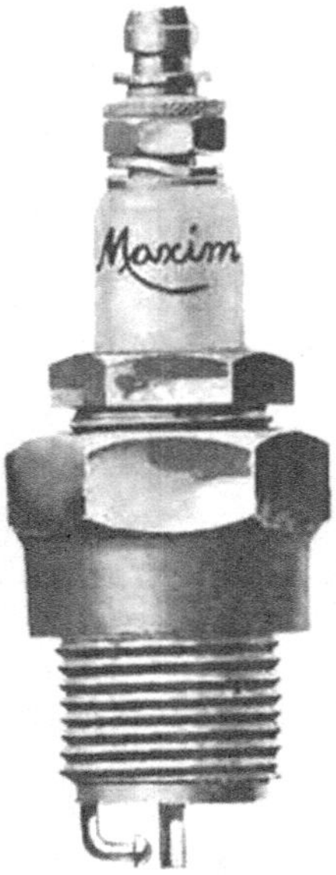

No. 1491—A.L.A.M. (Three - in - One Terminal). Price, each ......

Improved Design

Three-in-One Terminals

Accurate as a Time Fuse

No. 1494—A.L.A.M., Long (Three-in-One Terminal). Price, each.

No. 1501 — Special Chevrolet A.L.A.M., Long (special Chevrolet Terminal). Price, each ....

(Each Spark Plug Packed in a Neat Carton)

GUIDED by our previous experience and by a careful study of the principal Spark Plugs on the market, the design has been improved in several important respects. It is no exaggeration to say that we now consider the Maxim by far the best Spark Plug yet produced.

The Three-in-One Terminal does 50% more work than any other, and 200% more than most terminals. It takes the place of a slip terminal, a Buick and a screw connection terminal. It will therefore fit practically any make of Motor Car sold in Canada. The advantage of this to both dealer and motorist is apparent.

A Special Chevrolet Model is also supplied.

The heavy **Ball Point** which has always been a feature of higher grade Maxim Spark Plugs has been retained. Experience has proven that the Ball Point gathers a heavy pressure of current and produces a strong, rich spark, about **50% more powerful** than is obtainable by the use of other styles of points.

An interesting experiment to prove this to your satisfaction is to hold a number of thin sheets of paper in the course of a spark. The Maxim Spark will burn through twelve sheets, as against eight sheets penetrated by the spark of any ordinary plug.

"Accurate as a Time Fuse" is literally true of the new Maxim Spark Plug. It is made by mechanics who have been accustomed to working to a variation of not over .002, or 2-1000 of an inch, in making the delicate mechanism of a time fuse. Automatic machines of equal accuracy are used, and the designs, gauges and tools are similarly exact. As a result, all parts are interchangeable.

Allowance has been made in the design for the expansion of the various parts, for variation in the thickness of the porcelains, etc., and, in short, nothing has been neglected to insure the giving of the very best service under all conditions.

Seeing is believing, however, and we suggest that you try this improved Spark Plug for yourself. Our tests have convinced us that it is an article which it will pay dealers well to push, and which they can safely recommend.

No. 1495. Regular

Porcelains for Maxim Spark Plug

Only two models are now required.

No. 1495—Regular Porcelain, with Three-in-One Terminal ......Price, each
No. 1502—Special Porcelain, with Chevrolet Terminal ....Price, each

No. 1502. Chevrolet.

INDEX

T

V

W

ISBN #978-1-935700-52-4

Made in United States
Cleveland, OH
18 December 2025

28957062R00050